Managing Risk in Globalized Supply Chains

CAOLIONN O'CONNELL, ELIZABETH HASTINGS ROER, RICK EDEN,
SPENCER PFEIFER, YULIYA SHOKH, LAUREN A. MAYER, JAKE MCKEON,
JARED MONDSCHEIN, PHILLIP CARTER, VICTORIA A. GREENFIELD,
MARK ASHBY

Prepared for the Department of the Air Force
Approved for public release; distribution unlimited

RAND PROJECT AIR FORCE

For more information on this publication, visit www.rand.org/t/RRA425-1

Library of Congress Cataloging-in-Publication Data is available for this publication.
ISBN: 978-1-9774-0658-3

Published by the RAND Corporation, Santa Monica, Calif.
© Copyright 2021 RAND Corporation
RAND® is a registered trademark.

Support RAND
Make a tax-deductible charitable contribution at
www.rand.org/giving/contribute

www.rand.org

Preface

Given the likely exposure of U.S. Air Force (USAF) weapon system supply chains to foreign interests, USAF could potentially benefit from a robust supply chain risk management (SCRM) capability to evaluate and mitigate risks associated with the globalization of supply chains. To help USAF evolve its SCRM practices to keep pace with challenges of globalization, we researched academic literature on SCRM, reviewed federal and U.S. Department of Defense (DoD), and USAF policy and regulations related to supply chain management and acquisitions, and interviewed personnel from across USAF and DoD. This report recommends specific ways in which USAF can evolve its organization, policy, training, and data practices to avoid and to mitigate supply chain risk. This report is intended to help USAF approach SCRM from an enterprise perspective; a companion report offers practical guidance for those responsible for implementing SCRM within program offices.

The research reported here was commissioned by the Deputy Assistant Secretary of the Air Force for Logistics and Product Support (SAF/AQD) and conducted within the Resource Management Program of RAND Project AIR FORCE as part of a fiscal year 2019 project, *Globalization of the Supply Chain*. This report should be of interest to acquisition professionals and those responsible for mission assurance.

RAND Project AIR FORCE (PAF), a division of the RAND Corporation, is the Department of the Air Force's (DAF's) federally funded research and development center for studies and analyses. PAF provides the DAF with independent analyses of policy alternatives affecting the development, employment, combat readiness, and support of current and future air, space, and cyber forces. Research is conducted in four programs: Strategy and Doctrine; Force Modernization and Employment; Manpower, Personnel, and Training; and Resource Management. The research reported here was prepared under contract FA7014-16-D-1000.

Additional information about PAF is available on our website:
www.rand.org/paf/

This report documents work originally shared with the Department of the Air Force on September 23, 2019. The draft report, issued on September 27, 2019, was reviewed by formal peer reviewers and DAF subject-matter experts.

Contents

Figures and Tables

Figures

Tables

Summary

Issue

In recent years, policymakers have increased emphasis on national security risks deriving from globalization of weapon system supply chains to include foreign suppliers. We define *supply chain* as the tiered network of entities providing goods and services culminating in a deliverable to the U.S. Air Force (USAF), such as a product (e.g., a component of a weapon system) or a service. Suppliers are chiefly commercial firms and include manufacturers, service providers, and distributors. Inputs from upstream suppliers include raw materials, parts or components, machinery, and labor. Each supplier in the chain undertakes some production activity and then provides outputs to the next firm in the chain. The chain ends with the suppliers that are the prime contractors for USAF.

USAF could potentially benefit from a robust supply chain risk management (SCRM) capability to evaluate and mitigate risks.

Approach

To help USAF improve its SCRM practices, we researched academic literature on SCRM to identify common practices in the commercial sector; reviewed federal, U.S. Department of Defense (DoD), and USAF policy and regulations related to supply chain management and acquisition to understand existing guidance; and interviewed personnel from across USAF and DoD to learn about current practices and challenges.

Conclusions

Many opportunities exist for USAF to improve its management of supply chain risks, including better understanding the supply chain of its weapon systems, improving its own resilience to supply chain disruptions, and requiring and incentivizing its prime contractors and other suppliers to improve their own SCRM:

- USAF policies and responsibilities for SCRM are widely dispersed, with limited alignment, coordination, and information-sharing.
- USAF places implicit and explicit trust in its vendors to manage their supply chain risks sufficiently to protect USAF from the effects of disruptions.
- SCRM analysts have a limited understanding of which suppliers are actually in the supply chain of a given program or weapon system beyond the first tier.
- Although many sources of SCRM-relevant data exist, they are in diverse formats and can be difficult to access, integrate, and analyze.

- The USAF acquisition workforce of 2020 has had limited on-the-job training and exposure to the full range of supply chain risks and how to collect sufficient information to understand them.

Recommendations

This report recommends specific ways in which USAF can evolve its organization, policy, training, and data practices to avoid and to mitigate the effects of supply chain risk.

- USAF, and DoD more broadly, may benefit from an executive SCRM council to establish policy, set standards, and facilitate information-sharing across the enterprise. At a minimum, council membership would represent the acquisition, logistics, intelligence, counterintelligence, and operational user communities. A separate analytic organization within USAF could support the council's agenda.
- USAF may consider having its acquisition policy require programs that are strategically significant to consider supply risk as part of source selection.
- To gain insight into the lower-tier providers, some program offices could collect a complete list of raw materials, parts, and subcomponents and the associated suppliers needed to produce an end item (e.g., the bill of materials). To help prioritize SCRM activities, these program offices could also consider the value of requiring contractors to provide lists of items that can critically affect the reliability of contract end items (e.g., a critical items list) from contractors.
- To reduce the burden on program offices and to facilitate supply chain risk assessment, DoD could develop a comprehensive plan to manage SCRM-relevant data collected throughout government that would include USAF logistics, maintenance, and safety programs. In the absence of a DoD-wide effort, USAF may wish to consider developing a resource for its own community.
- DoD may benefit from developing an ongoing, formal, enterprise-level SCRM curriculum that trains acquisition professionals on identifying and mitigating supply chain risk.

Acknowledgments

The research question addressed in this report was conceived by Lawrence Kingsley, who was then Deputy Assistant Secretary for Logistics and Product Support. We are grateful for his foresight in sponsoring this project. We are grateful to Angie Tymofichuk, the current Deputy Assistant Secretary for Logistics and Product Support, for providing continued support for this project. We would also like to acknowledge John Ray and Robyn Downer of the Office of the Deputy Assistant Secretary of the Air Force for Logistics and Product Support. We are especially indebted to Ms. Downer, who provided assistance in connecting the research team with all the relevant organizations that conduct supply chain risk management across the Department of Defense.

We are grateful for all the individuals across the Department of Defense who spent a significant time discussing supply chain risk management with us. Their perspective and recommendations for improvement influenced our findings.

We would like to express our gratitude to all the RAND Corporation colleagues who provided valuable insight, guidance, and feedback. Thank you to Nancy Moore, Dara Gold, Jonathan Welburn, Don Snyder, Kenneth Girardini, Henry Willis, Sean Barnett, John Drew, Gabriel Lesnick, Daniel Schwam, Norah Griffin, Christopher Gilmore, Suzanne Genc, and Bradley Knopp.

Finally, we would like to thank our reviewers—Peter Whitehead, Michael Linick, and Frank Camm—who provided constructive feedback and helped improve clarity throughout. Naturally, any errors or omissions are solely the fault of the authors.

Abbreviations

AF/A4	Deputy Chief of Staff for Logistics, Engineering and Force Protection
AFMC	Air Force Materiel Command
AFOSI	Air Force Office of Special Investigations
AFI	Air Force Instruction
AFPAM	Air Force Pamphlet
AFPD	Air Force Policy Directive
ASM	approximate string matching
AVIC	Aviation Industry Corporation of China
BOM	bill of materials
CAGE	Commercial and Government Entity
CFIUS	Committee on Foreign Investment in the United States
CPU	central processing unit
DAF	Department of the Air Force
DBA	doing business as
DFARS	Defense Federal Acquisition Regulation Supplement
DIA	Defense Intelligence Agency
DLA	Defense Logistics Agency
DoD	U.S. Department of Defense
DoDI	Department of Defense Instruction
DoDM	Department of Defense Manual
DOJ	Department of Justice
DUNS	Data Universal Numbering System
EDGAR	Electronic Data Gathering, Analysis, and Retrieval
FAR	Federal Acquisition Regulation
FCPA	Foreign Corrupt Practices Act
FFATA	Federal Funding Accountability and Transparency Act

FIRRMA	Foreign Investment Risk Review Modernization Act of 2018
FPDS	Federal Procurement Data System
FPGA	field-programmable gate array
FY	fiscal year
GAO	Government Accountability Office
GIDEP	Government-Industry Data Exchange Program
IC	integrated circuit
ICD	Intelligence Community Directive
IPB	illustrated parts breakdown
ITAR	International Traffic in Arms Regulations
JWICS	Joint Worldwide Intelligence Communications System
MAJCOM	major command
MICAP	Mission Impaired Capability Awaiting Parts
NAVAIR	Naval Air Systems Command
NAVSEA	Naval Sea Systems Command
NDAA	National Defense Authorization Act
NIIN	National Item Identification Numbers
NIST	National Institute of Standards and Technology
NSN	National Stock Number
PDF	portable document format
RSP	readiness spares package
SAF/AQ	Assistant Secretary of the Air Force for Acquisition, Technology, and Logistics
SAF/AQD	Deputy Assistant Secretary of the Air Force for Logistics and Product Support
SAF/AQX	Deputy Assistant Secretary of the Air Force for Acquisition Integration
SCRM	supply chain risk management
SCRM TAC	Supply Chain Risk Management Threat Advisory Center
SCRM WG	Supply Chain Risk Management Working Group
SEC	U.S. Securities and Exchange Commission

SIPRNet	Secret Internet Protocol Router Network
SME	subject-matter expert
TSN	trusted systems and networks
USAF	U.S. Air Force
U.S.C.	U.S. Code

1. Introduction

Supply chains form the foundation on which all U.S. Air Force (USAF) operations depend. In recent years, Congress has increased emphasis on national security risks deriving from weapon system supply chains, with a particular emphasis on the globalization of the supply chain. For instance, the Fiscal Year (FY) 2018 National Defense Authorization Act (NDAA) banned the use of products and services from Kaspersky and prohibited the purchase of commercial equipment that supports nuclear command, control, and communications systems from several Chinese firms, including Huawei Technologies Company and ZTE Corporation.[1] Congress passed two additional pieces of legislation addressing avenues by which global supply chains could compromise national security. The Foreign Investment Risk Review Modernization Act of 2018 (FIRRMA) formalized and bolstered oversight of foreign investment in the United States under the interagency Committee on Foreign Investment in the United States (CFIUS).[2] FIRRMA was motivated by policymaker concerns that "the national security landscape has shifted in recent years, and so has the nature of the investments that pose the greatest potential risk to national security."[3]

Yet despite the risks of globalized weapon system supply chains, it is neither feasible nor desirable to eliminate all foreign sources from weapon system supply chains. Representatives of two organizations that conduct supply chain risk assessments suggested to us that, in the case of microelectronics, the U.S. Department of Defense (DoD) must accept the risk of foreign influence in the supply chain because it is economically challenging to meet DoD requirements through redomiciled production.[4] Moreover, access to foreign sources of supply has benefits beyond reduced acquisition costs; foreign sourcing can help mitigate some forms of supply chain risk. For example, if U.S. sources of supply face disruptions from natural disasters, the ability to turn to foreign alternatives reduces risk of disrupted supply. Given the inevitability of the exposure of USAF weapon system supply chains to foreign interests and the benefits of access to foreign sources of supply, USAF could potentially benefit from a robust supply chain risk management (SCRM) capability to evaluate and mitigate risks, including those associated with the globalization of supply chains.

---

[1] Pub. L. 115-91, National Defense Authorization Act for Fiscal Year 2018, December 12, 2017.

[2] FIRRMA is Title XVII of Pub. L. 115-232, John S. McCain National Defense Authorization Act for Fiscal Year 2019, August 13, 2018.

[3] Pub. L. 115-232, 2018.

[4] Interviews with Air Force Office of Special Investigations personnel, February 5, 2019, and with the Joint Federated Assurance Center, May 2, 2019.

Research Objective and Approach

The Deputy Assistant Secretary of the Air Force for Logistics and Product Support (SAF/AQD) asked RAND Project AIR FORCE to develop a repeatable SCRM process for the USAF to quickly identify, assess, prioritize, and mitigate supply chain risks that affect weapon system readiness.

We made several assumptions and limitations as the result of this direction. The key assumption is that the benefits of employing a SCRM process outweigh its associated costs. Our analysis did not assess the cost-effectiveness of the various recommendations within this report. In it, we discuss several systemic challenges that would prevent USAF from adopting a formal SCRM process. Addressing these challenges will require resources, and USAF will need to determine whether the potential benefits of the changes are an effective use of resources.

Our recommendations are based primarily on two sources. First, we compared existing and alternative practices against current practices from other DoD entities and industry. Where USAF practices deviate from these practices, we assessed the extent to which the differences stem from differing contexts or objectives—in which case, USAF deviations might be justified. If not, we recommend that USAF consider adopting (perhaps in a modified manner) the current practices. Second, we relied on information from interviews with SCRM practitioners and stakeholders across DoD—including program managers; intelligence community members; and logistics, acquisitions, and contracting personnel—regarding relevant SCRM challenges; SCRM practices that have been helpful or unhelpful; and what resources, organizational constructs, and policies the interviewees identified as necessary to achieve more effective outcomes.

The key limitation of a formal SCRM process is that it is useful only to the extent that the experts using it have been effectively trained. To address this limitation, a companion piece—a practitioner's guidebook—will offer practical guidance for those responsible for SCRM. Additionally, we recommend that DoD develop an ongoing, formal, enterprise-level SCRM curriculum that trains acquisition professionals on identifying and mitigating supply chain risk.

Because of the interconnectedness of the supply chain, many SCRM efforts are more appropriately executed at a level higher than USAF (e.g., U.S. government or DoD). In the absence of those efforts, USAF could forge ahead and develop SCRM capabilities that might be more cost-effectively done at a higher echelon. This report is intended to help USAF approach SCRM proactively from an enterprise perspective. Where applicable, we note recommendations that could potentially be better executed across DoD, but USAF may wish to consider creating an organic capability in the interim.

We used several research methods to conduct this research. To identify best practices in SCRM, we reviewed academic literature. To understand existing guidance, we reviewed federal, DoD, and USAF guidance related to supply chain management and acquisitions. To learn about current practices and challenges, we interviewed personnel from across USAF and DoD.

Organization of This Report

In this report, we focus on some of the systemic challenges that might prevent USAF from adopting a formal SCRM process. Given these challenges, we then recommend specific ways DoD and USAF can evolve organization, policy, training, and data practices to avoid and mitigate the effects of supply chain risk.

Chapter 2 is an assessment of how USAF is organized to implement SCRM and offers recommendations to improve organizational alignment with SCRM goals. In Chapter 3, we discuss incentivizing and enforcing SCRM for programs that potentially merit extra attention. Chapter 4 addresses the potential need for additional training of SCRM professionals in the face of new challenges. Chapters 5 and 6 address the collection, management, and analysis of data—both USAF and other—to support SCRM. Chapter 7 focuses on our major recommendations.

The appendix provides methodological details.

2. Organization and Policy for Supply Chain Risk Management

No single organization at the DoD or Air Force level is responsible for directing an enterprisewide SCRM vision. Perhaps as a result, no enterprise-level SCRM policy exists in either domain. Despite many ongoing DoD and USAF SCRM efforts, policy and guidance that explicitly define SCRM roles, responsibilities, and procedures are currently limited. This lack of guidance is unlikely to continue because Congress has already begun to pass legislation to address these shortcomings. Pub. L. 115–390 established a cross-agency, high-level Federal Acquisition Security Council, headed by the Office of Management and Budget, with responsibility and expansive authority for improving the security of supply chains for covered information and communications technology articles.[5] The council had its first meeting in March 2019, but it remains unclear what policy the council might enact. Additionally, the Air Force Security Enterprise Executive Board—a high-ranking forum within the Air Force tasked with providing security enterprise governance—established the SCRM Working Group (WG) in October 2018.[6] The SCRM WG charter is to establish and describe the mission, functions, construct, and responsibilities for SCRM governance.

Fragmented U.S. Air Force and Department of Defense Supply Chain Risk Management Policies

Air Force Policy Directive (AFPD) 23-1, *Supply Chain Materiel Management*, does give the Assistant Secretary of the Air Force for Acquisition, Technology, and Logistics (SAF/AQ) responsibility for developing a SCRM strategy, as well as promoting resiliency in supply chain sourcing and acquisition decisions.[7] However, according to the draft SCRM Campaign Plan, the development of a SCRM strategy is ongoing.[8] Several DoD organizations have partial authority and responsibility for SCRM, as specified in DoD Instruction (DoDI) 4140.01, *DoD Supply Chain Materiel Management Policy*,[9] and DoDI 5200.44, *Protection of Mission Critical*

---

[5] U.S. Code, Title 41, Public Contracts, Section 1322, Federal Acquisition Security Council Establishment and Membership; Pub. L. 15-390, Strengthening and Enhancing Cyber-Capabilities by Utilizing Risk Exposure Technology Act (SECURE Technology Act), December 21, 2018.

[6] SCRM WG, "Air Force Supply Chain Risk Management Working Group Charter," April 30, 2019.

[7] AFPD 23-1, *Supply Chain Materiel Management*, Washington, D.C.: Department of the Air Force, September 7, 2018.

[8] SCRM WG, "Air Force Supply Chain Risk Management Campaign Plan," undated.

[9] DoDI 4140.01, *DoD Supply Chain Materiel Management Policy*, Washington, D.C.: Under Secretary of Defense for Acquisition and Sustainment, March 6, 2019.

Functions to Achieve Trusted Systems and Networks (TSN),[10] but no overarching DoD lead exists.[11] These organizations include offices of the DoD Chief Information Officer, Under Secretary of Defense for Acquisition and Sustainment, Under Secretary of Defense for Research and Engineering, DoD component heads, and service secretaries.

Although not explicitly associated with SCRM, DoDI 5200.39, *Critical Program Information (CPI) Identification and Protection Within Research, Development, Test, and Evaluation (RDT&E)*, discusses the roles and responsibilities for identifying and protecting important capability elements of programs that merit protection, such as software algorithms, specific system hardware, training equipment, and maintenance support equipment.[12] This includes the intelligence and counterintelligence efforts supported by the Under Secretary of Defense for Intelligence; the Director, Defense Intelligence Agency (DIA); and the Director, Defense Security Service.

Current SCRM responsibilities for DoD and USAF appear to be mostly inferred from materiel management policy and guidance. The documents include DoDI 4140.01, the 12 volumes of DoD Manual (DoDM) 4140.01,[13] AFPD 23-1, and Air Force Instruction (AFI) 23-101, *Air Force Materiel Management*.[14] AFI 23-101, for example, authorizes the Deputy Chief of Staff for Logistics, Engineering and Force Protection (AF/A4) to provide implementing materiel management guidance, exercise enterprise oversight of materiel management, and define and collect supply chain management metrics from the major commands (MAJCOMs). As the materiel management lead, Air Force Materiel Command (AFMC) is responsible for overseeing centralized execution of enterprise materiel management operations and for tracking and analyzing supply chain management metrics. How these authorities and responsibilities relate to SAF/AQ's responsibilities for developing a SCRM strategy is unclear, because SCRM activities are characterized as a subset of supply chain management activities and very little guidance exists on SCRM coordination between SAF/AQ, AF/A4, and AFMC.

With no overarching SCRM policy or guidance for governance, SCRM-related activities that are specific to sources of supply chain risk are dispersed across different policies. DoD has separate policies and processes for acquisition programs, the management of counterfeit parts,

---

[10] DoDI 5200.44, *Protection of Mission Critical Functions to Achieve Trusted Systems and Networks (TSN)*, Washington, D.C.: DoD CIO/USD(R&E), July 27, 2017, change 3, October 15, 2018.

[11] Kerry R. McCarthy, Matthew R. Peterson, Jennifer J. Shafer, Jennifer Bisceglie, Dan Colman, and Brent Wildasin, *DoD Supply Chain Risk Management: Assessment and Recommendations: Assessment and Recommendations*, Tysons, Va.: LMI, March 2018.

[12] DoDI 5200.39, *Critical Program Information (CPI) Identification and Protection Within Research, Development, Test, and Evaluation (RDT&E)*, Washington, D.C.: Under Secretary of Defense for Intelligence and Under Secretary of Defense for Research and Engineering, May 28, 2015, change 2, October 15, 2018.

[13] DoDM 4140.01, *DoD Supply Chain Materiel Management Procedures* (in 12 volumes), various dates.

[14] AFI 23-101, *Air Force Materiel Management*, Washington, D.C.: Department of the Air Force, September 9, 2019.

TSN, and cybersecurity, which are all under the umbrella of SCRM, but DoD tackles them separately in its policy to achieve specific intents. Table 2.1 lists these policies and their various alignments. The linkages between these policies are unclear, potentially creating difficulties with coordinating the myriad of supply chain risks within an organization.[15]

Table 2.1. Multiple Policies Govern Aspects of Supply Chain Risk Management

SCRM-Related Activity	Policies
Program-related	• DoDI 5200.39, *Critical Program Information (CPI) Identification and Protection Within Research, Development, Test, and Evaluation (RDT&E)* • AFI 63-101/20-101, *Integrated Life Cycle Management*, Washington, D.C.: Department of the Air Force, March 7, 2013 • Air Force Pamphlet (AFPAM) 63-113, *Program Protection Planning for Life Cycle Management,* Washington, D.C.: Department of the Air Force, October 17, 2013
Counterfeit parts	• DoDI 4140.01, *DoD Supply Chain Materiel Management Policy* • DoDI 4140.67, *DoD Counterfeit Prevention Policy*, USD(A&S) April 26, 2013, change 3, March 6, 2020 • AFPD 23-1, *Supply Chain Materiel Management* • AFI 23-101, *Air Force Materiel Management* • AFPAM 63-113, *Program Protection Planning for Life Cycle Management*
Trusted systems and networks	• DoDI 5200.44, *Protection of Mission Critical Functions to Achieve Trusted Systems and Networks (TSN)* • AFI 63-101/20-101, *Integrated Life Cycle Management* • AFPAM 63-113, *Program Protection Planning for Life Cycle Management*
Cybersecurity	• DoDI 8500.01, *Cybersecurity*, Washington, D.C.: Department of Defense Chief Information Officer, March 14, 2014, change 1, October 7, 2019 • AFI 18-130, *Cybersecurity Program Management*, Washington, D.C.: Department of the Air Force, February 12, 2020

To understand how these separate endeavors may cause difficulty for coordination efforts, consider Air Force–level management of supply chain risks associated with untrusted systems and networks and counterfeit parts. TSN efforts pertain to the security of networks, and one threat to network security is the introduction of counterfeit microelectronics into the supply chain. TSN activities tend to be centralized within higher levels of the Air Force, while activities to combat counterfeit parts are more decentralized within lower levels. SAF/AQ is the Headquarters Air Force TSN focal point, performing enterprise TSN activities and coordinating with MAJCOMs on TSN requirements, best practices, and mitigations. AFMC acts in an intermediate coordination role for TSN by coordinating and prioritizing TSN resources, threats, etc., for program managers. An enterprise- or intermediate-level coordination role does not exist

---

[15] For example, McCarthy et al., 2018, uncovers such conditions specifically in the context of Operations and Sustainment (O&S).

6

for counterfeit parts. SAF/AQ develops strategies to guard against counterfeit parts in the supply chain and performance metrics for related programs but does not play a coordinating role in these activities. AF/A4, which has no role in TSN activities, establishes implementing policy on counterfeit parts. AFMC has responsibility to document and report appearances of counterfeit parts to Headquarters Air Force and users but does not have responsibility for coordinating and prioritizing such related aspects as mitigations. In fact, program managers report counterfeit parts to their MAJCOM TSN focal points, and that is the only link documented in policy or guidance between TSN and counterfeit parts.

Commercial Lessons About Organizing for Supply Chain Risk Management

For a comparison with DoD's implementation of SCRM, we examined how the commercial sector employs SCRM. In 2013, Deloitte surveyed 600 executives in manufacturing and retail companies to understand how the commercial sector organizes for SCRM.[16] Of these executives, 63 percent reported that their companies had risk management programs focused on supply chain; notably, nearly 60 percent had an upper-level executive accountable for supply chain risk.

Such commercial SCRM practices have been tested several times in real life, and USAF may gather lessons from these experiences. One of the most notable cases involves how Japanese car manufacturers responded to the March 11, 2011, Tōhoku earthquake and the ensuing Fukushima nuclear disaster.[17] Observers noted that Nissan managed to recover relatively quickly after the disaster, despite damage to six production facilities and problems with 50 of its critical suppliers, ascribing the company's recovery to a strong risk-management corporate culture. Notably, the company established a dedicated risk-management function, assigned individuals to manage specific risks, and an executive-level committee that regularly reported to the board of directors.

National Institute of Standards and Technology (NIST) work suggests that use of an executive-level SCRM is a common commercial practice. In 2014, NIST published a *Roadmap for Improving Critical Infrastructure Cybersecurity*.[18] The roadmap identified SCRM as an area for future focus. As a result, NIST researched current industry practice for cyber SCRM.[19] Although the focus was primarily on cyber SCRM, the 18 case studies discuss SCRM more broadly. This section summarizes the trends from those case studies, which included manufacturers, defense contractors, and information technology businesses.

The case studies indicate that many industries have two levels for conducting SCRM: executive and execution.

---

[16] Kelly Marchese and Siva Paramasivam, "The Ripple Effect: How Manufacturing and Retail Executives View the Growing Challenge of Supply Chain Risk," New York: Deloitte Development LLC, 2013.

[17] William Schmidt and David Simchi-Levi, "Nissan Motor Company LTC.: Building Operational Resiliency," Cambridge, Mass.: MIT Sloan Management, August 27, 2013.

[18] NIST, "NIST Roadmap for Improving Critical Infrastructure Cybersecurity," February 12, 2014.

[19] NIST, Cyber Supply Chain Risk Management," June 22, 2020.

The executive level is usually in the form of a panel or council composed of key stakeholders from across the company. The council develops basic policy and process guidelines, establishes consistency of practice, and shares information across the company. The council usually meets annually or quarterly and presents risks to senior management. *Juran's Quality Handbook*, a well-known industrial quality management reference that encompasses many of the same principles as SCRM, details a similar organizational practice under a quality leadership council.[20]

At the execution level, there is more variation of practice, in particular in how SCRM is aligned within the assorted companies. For example, Northrop Grumman aligns SCRM activities by product line; Procter & Gamble aligns them according to business unit; and Intel aligns them by regional focus. Additionally, companies employ both decentralized and centralized organizational approaches.[21] In general, decentralized execution is usually organized at the program level, and the organizations are simply given overarching guidance from the executive committee. A centralized organization involves a dedicated SCRM team that is the touchpoint between many business units and allows collaboration. As an example, multiplant organizations may have centralized responsibility for supplier quality on a product basis, meaning that a single plant is responsible for supplier quality on all purchases of product X, and that plant becomes the single coordinator for all other plants that buy product X.[22]

The private sector relies on a variety of mechanisms to support SCRM. There appears to be considerable emphasis on continuity—with the business both handling its own incident management and ensuring that its suppliers have their own business continuity plans in the event of a natural disaster or other significant disruption. Choosing the right supplier partnership is also an important component of SCRM in the private sector. *Juran's Quality Control Handbook* details the primary activities necessary for ensuring that products meet requirements with a minimum of inspections and corrective actions:[23]

1. Define product and program quality requirements.
2. Evaluate alternative suppliers.
3. Select suppliers.
4. Conduct joint quality planning.
5. Cooperate with the supplier during execution of contract.
6. Obtain proof of conformance requirements.
7. Certify qualified suppliers.
8. Conduct quality improvement programs, as required.
9. Create and utilize supplier quality ratings.

---

[20] Joseph A. Defeo, ed., *Juran's Quality Handbook*, 7th ed., New York: McGraw-Hill, 2017, p. 277.

[21] NIST, 2020.

[22] J. M. Juran and Frank M. Gryna, eds., *Juran's Quality Control Handbook*, New York: McGraw-Hill, 1988, p. 15.8.

[23] Quoted from Juran and Gryna, 1988, p. 15.3.

Many of these activities were involved in the NIST case studies. The emphasis was on ensuring that the contract requirements for suppliers reflect the buyer's priorities and on conducting audits to independently confirm that the contractual obligations were being met. The audits were often done under the auspices of an established supplier surveillance program.[24] Additionally, companies employed supplier evaluations to assess their potential partners. Finally, many companies shared their long-term and stable projections with suppliers to allow them to better anticipate upcoming needs.

Lessons from the Commercial Sector Applied to the Department of Defense

When learning from the commercial sector, DoD must keep in mind that its concerns and expectations for SCRM are different. These differences include the following:

- *Attack type.* Attacks, such as intellectual property theft and corporate sabotage, can affect the commercial sector and DoD similarly.[25] However, DoD faces greater threats than commercial sector from some types of supply chain attacks, such as logic bombs or backdoors.[26] Given the differences, the commercial sector might be optimized to protect against certain attacks that might affect a business's bottom line, but this does not apply for attacks that affect DoD operations only after delivery.

- *Dynamism of demand.* The commercial sector must be responsive to fluctuations in trends and market. However, the supply chains that support weapon systems must be responsive and agile enough to support major combat operations and a demand signal that would affect an entire industry. Additionally, a program office must be robust to uncertainty in funding and the inherent volatility of federal budgets. This concern is particularly relevant for classes of weapon systems that share only a handful of lower-tier suppliers that would be unable to accommodate fluctuations in demand across all weapon systems.

- *Length of sustainment.* DoD sustains weapon systems for decades, which means that they are more vulnerable to counterfeit electronics and obsolescence than the commercial sector.

- *Insurance.* Commercial entities can often insure against risk in ways USAF cannot. The primary strategic objective of commercial entities is to create profit. Because supply chain problems compromise profit, the goal of supply chain risk mitigation is to protect profit—not necessarily to prevent a supply disruption for its own sake. For commercial entities, supply chain risk is fundamentally a financial risk. Consequently, firms can satisfactorily mitigate supply chain risk by insuring against the financial consequences of

---

[24] Juran and Gryna, 1988, p. 15.27.

[25] Office of the United States Trade Representative, *2018 Special 301 Report*, Washington, D.C., 2018; Office of Strategic Services, *Simple Sabotage Field Manual*, January 17, 1944.

[26] A *logic bomb* is malicious code intentionally inserted into a software system that will trigger when specified conditions are met, such as geofences or a particular date and time. A *backdoor* enables unauthorized access and allows the user to bypass security protocols.

disruptions. In contrast, supply chain problems compromise USAF missions, so the burden on mitigation strategies is much higher: Disruptions must be avoided or their effects mitigated. USAF generally cannot achieve its mitigation objectives by insuring against risk but may wish to evaluate the consequences of allowing firms in its supply chains to mitigate risk through insurance.

- *Contracting.* Compared to their commercial counterparts, USAF buyers typically face more restrictions in how they solicit and negotiate with potential vendors, which can limit USAF's use of some supply chain mitigation strategies. For instance, acquisition regulations usually require USAF to specify requirements at the time of solicitation, prohibit USAF from withholding unclassified information about requirements from the public realm, and restrict USAF's ability to categorically exclude sources without specific authority. All these are potentially valuable supply chain risk-mitigation strategies available in the commercial sector. Contracts issued under Other Transaction Authorities face fewer restrictions than those subject to the Federal Acquisition Regulation (FAR) but can be used only for a small subset of requirements.[27]

- *Mitigation strategies.* Some commercial-sector mitigation strategies are not accessible to USAF. For example, the commercial sector has a variety of mechanisms to respond to resource shortages that are not applicable to DoD. When floods in Thailand affected Western Digital's disk drive production, its competitor, Seagate, capitalized on the shortage and auctioned some disk drives to the highest bidder.[28] Similarly, when materiel shortages affect multiple weapon systems, providers of that materiel could auction their goods to ensure that whoever receives it values it most, but federal regulations would likely prevent such an auction from occurring. Additionally, the commercial sector has more flexibility than USAF in shifting funds to respond to supply chain problems.

Given these differences, USAF should be careful in applying lessons from the commercial sector. But despite those differences, there are likely still valuable lessons that are relevant to DoD.

Organizing the U.S. Air Force for Supply Chain Risk Management

The two levels of SCRM management frequently seen in the commercial sector appear to offer a relevant lesson for USAF. For the executive level, initial strides are being taken at the enterprise level with the Air Force Security Enterprise Executive Board's establishment of the SCRM WG. An empowered executive council would facilitate development of policy and process guidelines, common and consistent SCRM practices, and efficient and robust information-sharing across the enterprise. The companion guidebook will discuss organizational strategies at the execution level.

---

[27] For a more in-depth discussion of Other Transaction Authorities, see Lauren A. Mayer, Mark V. Arena, Frank Camm, Jonathan P. Wong, Gabriel Lesnick, Sarah Soliman, Edward Fernandez, Phillip Carter, and Gordon T. Lee, *Prototyping Using Other Transactions: Case Studies for the Acquisition Community,* Santa Monica, Calif.: RAND Corporation, RR-4417-AF, 2020.

[28] Yossi Sheffi, *The Power of Resilience: How the Best Companies Manage the Unexpected*, Cambridge, Mass.: MIT Press, 2015, p. 75.

An executive level, with broad oversight and broad visibility throughout USAF and that is supported by centralized analyses, may be a beneficial first step in implementing a formal SCRM process. USAF SCRM activities might benefit from an empowered executive level for several reasons. First, a centralized organization would be able to focus on enterprise concerns that affect many systems, while program offices can focus on their specific concerns. Second, a centralized organization would have need-to-know broadly applied to ensure information-sharing with the relevant program offices. Third, it is unlikely that individual program offices would be willing to invest in foundational analyses to understand larger macroeconomic trends that may affect SCRM.

Commonality Among U.S. Air Force Weapon Systems

There is commonality of critical parts among USAF weapon systems. Here, we define *critical parts* as those that can render a system inoperable until the part is repaired or replaced. Such parts are particularly important in the supply chain risk context because of their effects on mission readiness and success.

To determine the critical parts for USAF weapon systems, we combined data from readiness spares package (RSP) kit inventory lists and the Mission Impaired Capability Awaiting Parts (MICAP) database.[29] Decisions determining the contents of RSP kits represent a certain level of SCRM that likely includes some parts that would not render a system inoperable. To complement the RSP, data from MICAP may be skewed to parts with supply issues or parts that break infrequently and are therefore not on hand but whose lack would render a system not mission capable. Together, these data sets represent what we believe to be the best estimate for critical parts.

Using this methodology, we estimated the critical parts or National Item Identification Numbers (NIINs) for 70 Air Force aircraft and weapon systems that are currently fielded.[30] In total, the database contained nearly 165,000 unique parts. Figure 2.1 illustrates that more than 2,800 critical parts are common among six or more weapon systems. The figure represents 62 of the 70 weapon systems. We would expect the number of common critical parts to grow if the weapon systems of other services were also included. Additionally, this analysis considers only parts that have associated NIINs and does not capture any commonality beyond that, such as software code.

By considering how many weapon systems are affected by a given part, SCRM analysts can focus their efforts on a more manageable subset of parts that are likely to affect multiple weapon systems (and thus whose disruptions would be more consequential). By examining the variety of suppliers for those parts, further scoping can be done to focus only on ones with single suppliers.

---

[29] Per the Air Force Guidance Memorandum to AFI 23-101, 2019, RSPs are used to support deployments of USAF weapon systems. Authorizations are based entirely on formal wartime tasking in the War and Mobilization Plan.

[30] NIINs are unique nine-digit codes used to distinguish each part.

Figure 2.1. Commonality of Critical National Item Identification Numbers Across Weapon Systems

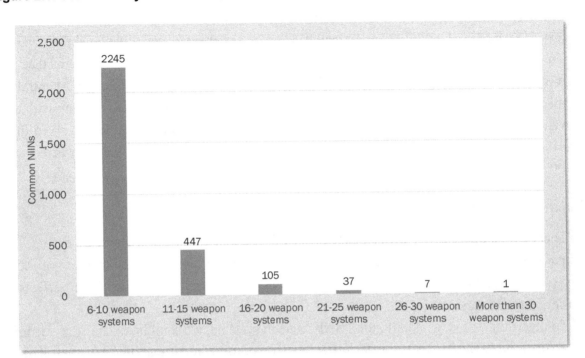

SOURCE: RAND analysis of MICAP and RSP data.

The commonality of critical parts across weapon systems indicates that these types of analyses should occur at a level that is sufficiently broad and that allows for trades among weapon systems. For example, a centralized organization can prioritize weapon systems that have higher strategic significance if resources are scarce for a particular common part. Finally, the commonality across weapon systems can enable mitigation strategies that would otherwise be infeasible for a single program office or unit to implement, such as contract consolidation to mitigate obsolescence.

Challenges of Information Sharing

Lack of information-sharing about the threat to supply chains and supplier risk assessments within DoD has been identified as a challenge for implementing SCRM routinely: "The lack of information-sharing between organizations means that the cross-cutting risk is not addressed at the higher level, leading to duplicative efforts and not leveraging best practices that have yielded results in another organization."[31] Having a named executive-level organization would provide a single point of contact for information flow, thereby limiting some duplicative efforts and ensuring that best practices are shared widely within the community. As an example, the Army has identified the Assistant Secretary of the Army for Acquisition, Logistics and Technology as the strategic manager of SCRM and the "focal point for processing requests to use other than

---

[31] McCarthy et al., 2018.

DoD intelligence components for producing weapon system intelligence mission data," which is a subset of the information-sharing challenges USAF faces.[32]

Additionally, "various parts of the Department of Defense . . . and the Intelligence Community . . . are generally aware of cyber and supply chain threats, but intra- and inter-government actions and knowledge are not fully coordinated or shared."[33] Addressing this challenge requires a larger DoD effort, but an executive-level organization at the Air Force level can at least potentially mitigate the duplication of efforts within USAF. Given the sensitivity of intelligence products, they are not widely disseminated and are shared only on a need-to-know basis. An entity with a broad perspective on SCRM activities within USAF could help identify those relevant parties faster.

Potential Need for Conducting Centralized Analyses for Supply Chain Risk Management

Personnel with responsibility for making SCRM decisions will likely be better off maintaining local analytic capability that allows decisionmakers to take advantage of local subject-matter expertise and ensure that they have the most responsive possible analytic support. However, some types of analysis can require technical skill or access to data that would not be easy or cost-effective to replicate across many units or programs. Specific conditions that make centralized data analysis more likely to make sense include analyses with one or more of following characteristics:

- The analysis requires information from multiple units or programs.
- The results would benefit multiple units or programs.
- Existing classification issues that would mean the required data or the analytic results might trigger further classification issues.
- There are policy implications that require centralized, cross-cutting, or complex responses
- The analysis requires specialized, hard-to-acquire analytic skills[34]
- necessitates establishing methodologies or benchmarks for program offices to execute.

As previously discussed, commonality among USAF weapon systems implies that collecting information from multiple programs and units provides some benefit. Building on this commonality, single program or units may not have sufficient data for anomaly detection, which increases with more data. The commonality of critical parts among weapon systems indicates

---

[32] Army Regulation 70–77, *Program Protection*, Washington, D.C.: Headquarters, Department of the Army, June 8, 2018.

[33] Chris Nissen, John Gronager, Robert Metzger, and Harvey Rishikof, *Deliver Uncompromised: A Strategy for Supply Chain Security and Resilience in Response to the Changing Character of War*, McLean, Va.: MITRE Center for Technology & National Security, August 2018.

[34] A specific example might include analyses that are not directly informed by readily available data, but which that involve making inferences based on available—but imperfectly informative—data.

that a centralized analysis might produce actionable intelligence faster.[35] Building on the analyses of individual common parts could lead to an analogous analysis that focuses on companies that are suppliers for multiple programs. Chapter 5 discusses the limited data currently available and value of collecting such data in future.

Classification issues underlie many information-sharing issues. Depending on the program, unit, or even installation, access to the appropriate network may be limited. Having a known centralized organization that has access to all classified networks and any data on them can provide reachback for organizations that do not.

Industrial-base analyses are relevant to SCRM and may have broad policy implications. For example, USAF's Office of Commercial Economic Analysis examines market-based threats across sectors and industries in the defense industrial base.[36] This relatively new office is currently in the process of expanding and developing its capabilities.

Currently, no single USAF organization is responsible for overseeing SCRM-specific analyses and guiding the development of methodologies to support such assessments. Centralized analysis of markets and economic conditions could be helpful for inferring the financial and demand stability of critical industrial sectors and for rapidly identifying changing market conditions. For example, understanding the likely financial conditions that precipitate a bankruptcy would help inform how acquisition professionals should review reports that are filed with the U.S. Securities and Exchange Commission (SEC) to ensure potential suppliers are not at risk of bankruptcy.

Recommendation: Improve U.S. Air Force Organization for Supply Chain Risk Management

Current executive management of the USAF supply chain focuses on only certain subsets of the supply chain (e.g., cyber, counterfeit, obsolescence). With the creation of the SCRM WG, strides are being taken toward a more comprehensive approach. USAF may benefit from an executive council that focuses on SCRM more broadly to establish policy, set standards for SCRM, and facilitate information-sharing across the organization. This executive council would benefit from an organization that is able to conduct centralized analyses that are beyond the scope of program offices. This organization can help facilitate information-sharing across USAF. Additionally, program offices that merit additional SCRM attention may consider creating supply chain working groups composed of, at a minimum, program office personnel, Defense Logistics Agency (DLA) representatives, supply chain management wing personnel, contracting and engineering personnel, and any other relevant stakeholders.

---

[35] David A. Galvan, Brett Hemenway, William Welser IV, and Dave Baiocchi, *Satellite Anomalies: Benefits of a Centralized Anomaly Database and Methods for Securely Sharing Information Among Satellite Operators*, Santa Monica, Calif.: RAND Corporation, RR-560-DARPA, 2014.

[36] Interviews at USAF Office of Commercial Economic Analysis, December 12, 2018.

3. Incentivizing and Enforcing Supply Chain Risk Management

Currently, DoD program managers are incentivized to meet cost, schedule, and performance goals. Existing policy and regulation are of limited utility in compelling SCRM, and most acquisition professionals assume the contractor is actively managing its supply chain. However, without clearer direction or financial incentives, the contractor is likely optimizing for cost, schedule, and performance, narrowly defining *performance* as reaching certain capability thresholds, not necessarily for security.

The Government Must Be Responsible for Supply Chain Risk Management

Acquisition professionals generally presume that it is in each contractor's best interest to manage its own supply chains. The Defense Federal Acquisition Regulation Supplement (DFARS) provides clauses that direct contractors to mitigate supply chain risk, but these clauses do not include an enforcement mechanism or, for that matter, a definition of *supply chain risk*. A MITRE report highlighted this as a structural challenge DoD faces: "Overreliance on 'trust,' in dealing with contractors, vendors, and service providers, has encouraged a compliance-oriented approach to security—doing just enough to meet the 'minimum' while doubting that sufficiency will ever be evaluated."[37]

Within DoD, DFARS Subpart 239.73 requires DoD program offices and contracting officers to manage supply chain risk for the "acquisition of information technology for covered systems." This subpart requires the addition of two clauses (DFARS 252.239-7017 and 252.239-7018) in all solicitations and contracts for information technology services or supplies, including commercial item acquisitions.[38] Paragraph (b) of DFARS 252.239-7018 states that "[t]he Contractor shall mitigate supply chain risk in the provision of supplies and services to the Government," but does not define what is meant by *mitigating risk*, and, assuming that a set of actions was defined to illustrate mitigation, it lacks compliance verification. Both of these DFARS clauses state that "to manage supply chain risk, the Government may . . . consider information, public and non-public, including all-source intelligence, relating to an offeror and its supply chain." However, these clauses do not provide the government with specific authorities to collect information from a prime contractor regarding its supply chain. Unless SCRM procedures are explicitly weighted in source selection and unless sufficiency criteria are defined, the contractor is unlikely to be incentivized. Again referencing MITRE's report:

---

[37] Nissen et al., 2018.

[38] Code of Federal Regulations, Title 48, Federal Acquisition Regulations System, Section 252.239-7017, Notice of Supply Chain Risk; Code of Federal Regulations, Title 48, Federal Acquisition Regulations System, Section 252.239-7018, Supply Chain Risk.

> All too often today, DIB [defense industrial base] contractors are reluctant to price added integrity and integrated risk management into their bids because the U.S. government rarely requires it in the Request for Proposal (RFP), and they fear losing the contract where higher cost may be a decisive negative discriminator.[39]

In contrast to the DFARS clauses, FAR 52.249-14 absolves suppliers of risks associated with natural disasters or with "acts of God or the public enemy" with a *force majeure* clause.[40] This clause transfers risks associated with a nation-state employing a supply chain attack from the supplier to USAF, assuming the supplier was not negligent. A prior RAND report quoted one aerospace-sector SCRM official: "One thing we look at is what is in the contract. If there is a force majeure clause, then the cost of realizing a supply chain disruption due to natural causes is passed on to the customer."[41] Such clauses mean that suppliers are not responsible for disruptions considered outside their control. Although usually associated with natural disasters, these disruptions can also include risk associated with a nation-state seeking to undermine the supply chain of a weapon system. Given this, the burden is on USAF to define specific actions that suppliers must take to protect the supply chain. USAF might not be able to hold vendors accountable for the outcome because of *force majeure* but can still hold them responsible for not abiding by clauses requiring certain actions or resource allocations, the absence of which might reduce vendor resilience to events falling under *force majeure*.

The Government Does Not Prioritize Security in Source Selection

Currently, DoD program managers are incentivized to have contractors meet cost, schedule, and performance goals. Current authorities allow program managers to include SCRM considerations in source selection, and the defense industrial base needs this incentive to justify investing in SCRM, but program managers lack the guidance or the requirement to do so.

Section 881 of the FY 2019 NDAA added a section titled Requirements for Information Relating to Supply Chain Risk to the U.S. Code (U.S.C.).[42] This statute gives certain agencies (including DoD) the authority to gather information regarding prospective contractors that may provide "covered systems" (such as information technology) and to exclude sources from being prime contractors or subcontractors that pose higher-than-acceptable levels of supply chain risk. However, this statute does not require program offices or contracting officers in DoD or its services to consider such risk, and defining acceptability of risk is also important.

---

[39] Nissen et al., 2018, p. 19.

[40] Code of Federal Regulations, Title 48, Federal Acquisition Regulations System, Section 52.249-14; Nancy Y. Moore and Elvira N. Loredo, *Identifying and Managing Air Force Sustainment Supply Chain Risks*, Santa Monica, Calif.: RAND Corporation, DB-649-AF, 2013.

[41] Moore and Loredo, 2013, p. 51.

[42] Pub. L. 115-232, 2018; U.S. Code, Title 10, Armed Forces, Section 2339a, Requirements for Information Relating to Supply Chain Risk.

The Competition in Contracting Act, as codified for DoD in 10 U.S.C. 2304, generally requires agencies to use competitive source selection for government contracts.[43] According to 10 U.S.C. 2305, agencies are required to "specify the agency's needs and solicit bids or proposals in a manner designed to achieve full and open competition for the procurement."[44] In general, however, agencies have broad latitude to determine the specifications, parameters, and evaluation factors they use in their solicitations, which "shall depend on the nature of the needs of the agency and the market available to satisfy such needs." Rather, the rule is that an agency must conduct a competition in a "full and open" manner that does not unfairly advantage one competitor over another, or else agencies may be subject to bid protests filed by disappointed bidders with the agency, the Government Accountability Office (GAO), or the Court of Federal Claims. These broad statutory authorities allow an agency to consider supply chain risk as a factor, but they fall short of requiring the consideration of this risk.

In general, Title 10's legal requirements for competition are implemented through FAR Part 15, and the agency promulgates regulations or guides that interpret and apply FAR Part 15. FAR Subpart 15.3 describes the procedures for source selection. Mirroring the language of 10 U.S.C. 2305, FAR 15.304 states that an agency's "award decision [shall be] based on evaluation factors and significant subfactors that are tailored to the acquisition."[45] Further, FAR 15.304(c) states that the "evaluation factors and significant subfactors that apply to an acquisition and their relative importance are within the broad discretion of agency acquisition officials." The FAR does not specify which factors must be considered, except to state that each solicitation should contain some consideration of price or cost, quality (i.e., "past performance, compliance with solicitation requirements, technical excellence, management capability, personnel qualifications, and prior experience"), and past performance. FAR Part 15 explicitly allows contracting agencies to consider "performance risk" and other types of risk in their source selection decisions, as they judge which proposals will provide the "best value" to the government, and to document the "relative strengths, deficiencies, significant weaknesses, and risks" associated with each proposal in the agency's contract file. The FAR 15.304(c) provides the authority needed to incorporate SCRM considerations in source selection, but this is not generally done under current practices.

Applying Supply Chain Risk Management Priorities Appropriately

Thus far, program offices define the requirements for each source selection in enough detail to make local decisions about awarding contracts, but issues related to SCRM are frequently not

---

[43] U.S.C., Title 10, Armed Forces, Section 2304, Contracts: Competition Requirements. The act itself is Title VII in Pub. L. 98-369, Deficit Reduction Act of 1984, July 18, 1984.

[44] U.S.C., Title 10, Armed Forces, Section 2305, Contracts: Planning, Solicitation, Evaluation, and Award Procedure.

[45] Code of Federal Regulations, Title 48, Federal Acquisition Regulations System, Section 15.304, Evaluation Factors and Significant Subfactors.

part of that calculus. In many cases, the costs associated with prioritizing security with respect to supply chains outweighs any potential benefits, but in some cases these costs are potentially worthwhile.

For example, the intelligence community and the Army direct additional SCRM activities under certain conditions. In these cases, it may be worthwhile to prioritize security and SCRM in the source-selection process, which would, in turn, support the SCRM activities after acquisition commences. The intelligence community directs a supply chain risk assessment of any acquisition item that is mission critical or is intended to be used in a classified user environment.[46] The heads of the intelligence community are tasked with identifying what is designated mission critical. Similarly, the Army directs SCRM for national security systems, automated tactical systems, and automated weapon systems, Mission Assurance Category I systems,[47] systems registered as mission critical, and other systems that the Army Acquisition Executive or Chief Information Officer/G–6 determines are critical to the direct fulfillment of military or intelligence missions.[48]

This prioritization scheme is also seen in the commercial sector. There is one example of electronics organization that separates its purchases into two categories: (1) items for which quality standards have been well established and (2) bespoke items. The quality will depend on the supplier chosen.[49] For purchases in the first category, purchasing managers are told to focus on the lowest cost; for purchases in the second category, managers are directed to select suppliers with the highest quality standard, even at a cost premium.

In reviewing Air Force policy, we did not find any direction to help with prioritizing which programs merited additional attention to SCRM. For example, AFI 23-101 effectively applies to "all government-owned property."[50] AFI 63-101/20-101 requires a program-protection plan for all acquisition category programs.[51] Similarly, many program-protection activities that could be categorized as SCRM are broadly applicable to all acquisition-category programs in AFPAM 63-113.[52] If USAF chooses to establish an empowered executive council for SCRM, that council could help identify which programs should be required to engage in comprehensive SCRM activities and provide guidance on how to prioritize security in the supply chain as part of the source-selection process.

---

[46] Intelligence Community Directive (ICD) 731, *Supply Chain Risk Management*, Office of the Director of National Intelligence, December 7, 2013; Intelligence Community Standard (ICS) 731-02, *Supply Chain Threat Assessments*, Office of the Director of National Intelligence, May 17, 2016.

[47] As defined by DoDI 5200.44, 2017.

[48] Army Regulation 70–77, 2018.

[49] Juran and Gryna, 1988, p. 15.6.

[50] AFI 23-101, 2019.

[51] AFI 63-101/20-101, 2013.

[52] AFPAM 63-113, 2013.

Neither the Air Force FAR Supplement nor Mandatory Procedure 5315.3 mandates or precludes the inclusion of supply chain risk as a source-selection factor.[53] Given that, additional FAR clauses are not necessary. Instead a program office will need guidance on how to evaluate and score what the different offerors provide in terms of SCRM and how to prioritize those capabilities over other competing demands. By including security and SCRM in the source-selection process, USAF indicates that it is prioritizing these considerations in this instance and signals the importance of SCRM to the contractor executing the eventual contract.

Recommendation: Consider Supply Chain Risk During Source Selection for Key Programs

USAF acquisition professionals may need to think beyond cost, schedule, and narrowly defined performance risk and should probably extend consideration to security. USAF, likely through the SCRM WG or some other executive council, could develop policy to identify which acquisition programs merit additional SCRM attention once acquisition commences. This policy could be modeled after existing guidance from the Army or intelligence community. For programs that meet these criteria, USAF should consider developing policy instructing program managers to consider security and risk management as part of the source-selection process.

---

[53] Mandatory Procedure 5315.3, "Source Selection," *Air Force Federal Acquisition Regulation Supplement,* 2019.

4. Supply Chain Risk Management Requires a Deep Knowledge of the System

Over the course of this analysis, we spoke with supply chain specialists from across DoD. Many emphasized that, as weapon systems grow in complexity and capability, the definition of *supplier* is becoming broader; as a result, risk assessments require extensive information. For example, USAF not only needs to know about a traditional supplier that provides a deliverable, but also nontraditional suppliers, such as the company that provides automation for the traditional supplier's manufacturing floor. To illustrate this phenomenon, we use a single field-programmable gate array (FPGA) as a case study.

Increasing Complexity of the Supply Chain

When we spoke with representatives of DoD organizations that conduct supply chain risk threat assessments, several subject-matter experts (SMEs) noted that program offices requesting these assessments often provided insufficient information to actually conduct an assessment.[54] These experts noted that requesters frequently lacked the technical knowledge necessary to understand what was required for a comprehensive assessment, and one mentioned that FPGAs illustrate his point. FPGAs are essentially reconfigurable digital circuits with no defined function at the time of manufacture. In contrast with the general-purpose processors (e.g., central processing units [CPUs]) found in laptops and mobile devices, FPGAs must be configured at power-up.[55] This versatility affords several advantages, and demand for FPGAs—or, more generally, *programmable logic*—has been on the rise.[56]

Figure 4.1 illustrates the key points in an FPGA life cycle.

---

[54] Interviews at the SCRM Threat Advisory Center (SCRM TAC), February 28, 2019, and the Naval Surface Warfare Center, Crane Division, June 10, 2019.

[55] FPGAs can be configured to emulate nearly any type of integrated circuit (IC). For this reason, they were developed primarily as a prototyping platform for CPUs or other high-volume production ICs.

[56] Stephen Trimberger, and Jason J. Moore. "FPGA Security: Motivations, Features, and Applications," *Proceedings of the IEEE*, Vol. 102, No. 8, August 2014; MarketWatch, "Global Embedded Field-Programmable Gate Array (FPGA) Market by 2023: Global Industry Report with Manufacturers, Regions, Trends, Challenges, Market Size, Product Types and Applications," press release, June 30, 2020.

Figure 4.1. Life Cycle of a Generic FPGA

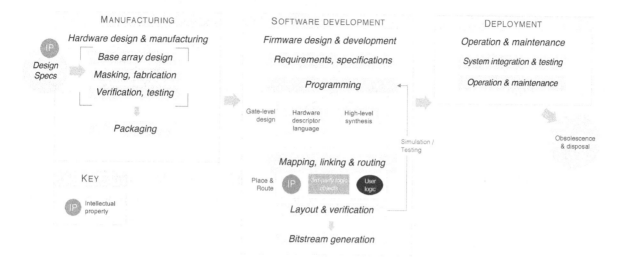

SOURCE: Derived from Raymond C. Shanahan, "Field Programmable Gate Array (FPGA) Assurance," presentation at the 20th Annual NDIA Systems Engineering Conference, Springfield, Va., October 26, 2017.

Manufacturing

FPGAs are designed and sold by several *fabless* chipmakers, meaning these chipmakers outsource fabrication to semiconductor foundries.[57] The basic workflow is as follows: All product requirements, design specifications, and intellectual property are delivered to the foundry from the designer. Various tools are then used to convert high-level instructions and design requirements into gate-level logic and circuitry to be impressed on a thin silicon wafer. After extensive testing, the completed wafer is encapsulated in a protective casing and sent to final assembly for circuit-board mounting. Most program offices can easily identify the designer and foundry for FPGAs, but that information provides an incomplete picture of an FPGA's supply chain.

Software Development

Once FPGAs have been manufactured, the hardware must be programmed and instructions provided at the gate level.[58] FPGAs have been notably difficult to program and operate, but tools have become available that allow common programming languages to be converted into

---

[57] Paul Dillen, "And the Winner of Best FPGA of 2016 Is . . . ," *EE Times*, June 3, 2017. Three companies control more than 90 percent of the FPGA market: Xilinx, Intel, and Microsemi (Dillen, 2017). Although multiple foundries are available, only a select few compete at the leading edge because fabrication is remarkably complicated, requiring large investments and immensely high-precision tools.

[58] A general processor (CPU) contains a fixed set of instructions that correlate with the hardware; high-level programming languages (C/C++, Fortran, etc.) use these instructions to delivery functionality.

integrated gate-level instructions.[59] As a result, the software development process now includes complex algorithms provided by a number of open-source libraries or proprietary software. These synthesis tools (often provided by the vendor) have become indispensable.

FPGAs are dependent on a limited number of toolkits, and inserting malware or corrupt code into the basic development tools or software libraries upstream could potentially infect millions of devices without detection.[60] Notably, USAF has contracted for the development of a private cloud-computing infrastructure to enable the secure use of IC design software tools.[61]

Deployment

After an FPGA has been programmed, it is ready for deployment. Once deployed, the FPGA is largely indistinguishable from other ICs and will operate nominally until power is removed, at which point the memory is wiped clean. New software can potentially be pushed to steadily improve efficiency and security.[62]

Disposal

After an FPGA has served its intended purpose or has become obsolete, proper device disposal is critical. As with many IC components, only complete destruction ensures that intellectual property and protected designs do not land in the wrong hands. As a complement to disposal at the end of the chip's life, high-volume fabrication at the initial manufacturing stage can lead to excess production (discards), which may also not be disposed of correctly.

Supply Chain Risk Management Implications of Increased Complexity

FPGAs provide an illustrative example of the challenges associated with modern IC solutions and the expansive web of suppliers that accompany them. In this case, the supply chain of an FPGA must include not only the designer and foundry but also their software tools and the software tools used in programming and maintenance. Additionally, consideration should be given to tracking disposal plans at both initial production and end of life to ensure appropriate risk exposure.

Beyond FPGAs, *suppliers* may need to be interpreted more broadly; for example, software installed to automate manufacturing facilities may also be relevant to SCRM. Given these issues, a recent MITRE report on supply chain security and resilience advocated that "[a]cquisition

---

[59] David F. Bacon, Rodric M. Rabbah, and Sunil Shukla, "FPGA Programming for the Masses," *Communications of the ACM*, Vol. 56, No. 4, April 2013. Historically, only two programming languages have existed for FPGAs: Verilog and VHDL; both are described as hardware descriptor languages and are known for their difficulty.

[60] Andy Greenberg, "Software Has a Serious Supply-Chain Security Problem," *Wired*, September 18, 2017.

[61] Nimbis Services Incorporated, "Trusted Silicon Stratus (TSS) Workshop," Mclean, Va., February 2011.

[62] With FPGAs, new software implies an entire system rewrite, not just a standard patch or update. This capability can substantially extend the device's operating lifespan and improve security late in the life cycle.

contract language should require the disclosure of commercial, open-source, and third-party software components as part of an SBOM [software bill of materials]. These disclosures should be independently verified."[63] That independent verification can be done by a third party (e.g., an auditing firm) or by the government customer.

The acquisition workforce is currently not informed about these risks and how to collect the full range of information to understand them. Bill Evanina, director of the National Counterintelligence and Security Center in the Office of the Director of National Intelligence, was quoted in an interview discussing the Department of the Navy's 2019 Cybersecurity Readiness Review as saying that the acquisition workforce is

> the least educated with respect to the counterintelligence and security threat. Their job is to acquire, procure and get things online as soon as possible. We have an obligation to advise and inform them on what those threats look like and provide them some tools to do some basic due diligence. . . . We have set out some standards to talk about what those due diligence standards are. If you [are] going to award a contract for a printer or a fax machine, just Google the company and make sure they exist. Let's just make sure they are a legitimate company.[64]

Several organizations that conduct risk assessments expressed the need for program offices to be better informed and trained on information that should be requested from contractors prior to conducting risk assessments.[65]

Recommendation: Train Acquisition Professionals on Supply Chain Risk

Acquisition professionals would benefit from understanding supply chain risks better and where to find data to help assess those risks and inform mitigation strategies. A companion guidebook will address this benefit, but it would behoove USAF to consider creating a curriculum or taking advantage of courses offered by other services. Such a curriculum would potentially be beneficial for such personnel as contracting officers, program managers and engineers, quality assurance managers, and logisticians.

U.S. Navy civilian personnel at Naval Surface Warfare Center–Crane and Naval Air Systems Command–Patuxent (NAVAIR-Patuxent) have developed and implemented SCRM training curricula that may serve as models for USAF curriculum development. Naval Surface Warfare Center–Crane SMEs provide training for DoD and industry employees focused on counterfeit materiel that includes information on how to implement Department of the Navy policies, how to identify risky suppliers, how to report counterfeits, how to authenticate and investigate materiel,

---

[63] Nissen et al., 2018.

[64] Jason Miller, "Why the Navy Is Giving Agencies, Industry a Much-Needed Wake-Up Call on Supply Chain Risks," Federal News Network website, April 4, 2019.

[65] Interviews at SCRM TAC, February 28, 2019, and the Naval Surface Warfare Center, Crane Division, June 10, 2019.

and how to dispose of or recycle materiel properly.[66] Personnel based at NAVAIR-Patuxent provide products and/or training for four different classes and will provide it to USAF personnel on request.[67] These courses focus on sharing relevant terms and definitions, insights into challenges that supply chain executives at prime contractors face and their motivations for using subcontractors, and common risks to the supply chain. Course participants are also provided with checklists of questions to ask suppliers to facilitate information-gathering. Industry trends, negotiation tactics, and supplier analysis tools are also discussed. Although USAF personnel could participate in this training, USAF leadership may want to consider developing a USAF-focused curriculum that incorporates unique USAF considerations and policy.

---

[66] Interview at Naval Surface Warfare Center, Crane Division, May 22, 2019.

[67] Interview at NAVAIR-Patuxent, June 13, 2019.

5. Limited Data Available to Understand Supply Chain

Understanding SCRM-relevant characteristics of companies in the supply chain is necessary to understand the risks that might be associated with these companies and how to appropriately manage the risks. Understanding the security-related characteristics of the firms in the first, second, third tiers of suppliers requires having a foundation of data.[68] This chapter discusses the limited data currently available to the government to understand the supply chain of specific weapon systems.

We recommend that program offices consider collecting a comprehensive list from their prime contractors of the raw materials, subassemblies, assemblies, and parts required to manufacture or repair an end item. This list is frequently referred to as a bill of materials (BOM) or, frequently, as an illustrated parts breakdown (IPB). To better understand subtier suppliers, portions of the BOM that are subcontracted would be noted and the suppliers identified. Initially, these requests can be limited to program offices identified by policy as meriting additional attention to SCRM.

Limitations on Mapping the Supply Chain

The federal government has a number of systems that contain data on prime contractors and subcontractors.

The System for Award Management and the Federal Procurement Data System (FPDS) provide insight into DoD's prime, or Tier 1, contractors. The System for Award Management registers all businesses wishing to do business with the federal government. Businesses must provide information on their annual revenue, numbers of employees, and the industries in which they seek to provide goods and services. FPDS contains contract actions for all federal purchases above the micropurchase threshold. FPDS–Next Generation contains the dollar value of the total award and the specific obligation for the action; industry and product and service codes for the goods and services being procured; and other contractor characteristics, including locations.

The Federal Funding Accountability and Transparency Act (FFATA) Subaward Reporting System provides Tier 2 supplier information and began collecting data in 2010. In accordance with FAR 52.204-10, this system reports subawards that are greater than $25,000 (exempting contractors with annual revenue of less than $300,000), but prior RAND research indicates that

---

[68] The prime contractor is Tier 1, and the first layer of subcontractors is Tier 2. Subcontractors can also have subcontractors, and these are Tier 3 suppliers. A complex weapon system can have many tiers of suppliers. This is frequently referred to as data-information-knowledge-wisdom (DIKW) hierarchy or pyramid. See also Russell L. Ackoff, "From Data to Wisdom," *Journal of Applied Systems Analysis*, Vol. 16, No. 1, 1989.

the data are incomplete.[69] Additionally, the high threshold means that a portion of the supply chain is not being captured.[70] Given these limitations, acquisition professionals have incomplete insight into their supply chains unless they specifically request this information in contracts. In turn, a contractor can only report what it knows, and commercial firms often have little insight more than two tiers down from themselves.

The NAVAIR recognized this issue as it was standing up its own SCRM process, and, over the past decade of implementation, has developed effective methods of gaining visibility into its supply chains.[71] As a result, NAVAIR began routinely requesting a BOM or IPB from contractors and, importantly, a commensurate list of any suppliers for those parts. When possible, it will request a BOM and list of suppliers for any major subcontractors, thereby getting insight into Tier 3 suppliers. Considerable effort is expended putting the various BOMs into a consistent format and into a database to enable analysis; the data are requested on a routine basis to ensure accuracy. To minimize this additional work, SMEs recommended making this requirement a data item description in the contract to ensure uniformity. NAVAIR has had success understanding the intersection of suppliers with different programs and has taken advantage of Defense Priorities and Allocations System ratings to prioritize parts at a single supplier that supports multiple programs.[72]

As programs move from acquisition to sustainment, issues related to obsolescence and diminishing manufacturing sources and material shortages become more frequent. When no other vendors are available, programs must consider replacements that comply with the form, fit, and function of the original part. This means that a program may wish to have ownership of its technical data and that the data would be most beneficial if comprehensive. We heard one example of having the technical drawings for one part but no information on the material that was used to create it.[73] Recent RAND research found a host of failures related to collecting technical data packages on weapon systems, for example, government personnel inappropriately acceding to contractor claims about what rights the government could acquire; government personnel acquiring the data rights but failing to list technical data as deliverables or failing to take delivery of technical data before relevant contract authority expired; and disputes arising

---

[69] Nancy Y. Moore, Clifford A. Grammich, and Judith D. Mele, *Findings from Existing Data on the Department of Defense Industrial Base*, Santa Monica, Calif.: RAND Corporation, RR-614-OSD, 2014.

[70] Using data from USA Spending that we pulled and analyzed January 4, 2020, roughly 10 percent of the prime contracts with DoD are below $25,000 between FYs 2008 and 2020 (USA Spending, undated a). Naively, we would assume that even more than 10 percent of subcontracts would be below $25,000 because FSRS captures a lower tier, but we are unable to estimate what portion of the supply chain is not being captured with that threshold.

[71] Interviews with current and former NAVAIR personnel, June 13, 2019, and July 16, 2019.

[72] Defense Contract Management Agency, "Defense Priorities & Allocations System (DPAS)," webpage, May 7, 2019.

[73] Interview at Minuteman III Program Office, January 18, 2019.

between the government and contractors over rights, usually triggered by the marking of data by contractors.[74]

In addition to technical data, program offices being able to routinely collect data on suppliers from the prime contractor could help acquisition professionals identify prior sources of supply for obsolete parts. Similarly, one interviewee mentioned using the Advanced Component Obsolescence Management system to identify parts that are common among weapon systems, such that multiple programs can benefit when a new vendor is qualified.[75]

Although collecting data can inform SCRM, analysis can be difficult due to the sheer number of firms. When Naval Sea Systems Command (NAVSEA) began collecting data on subcontractors from seven ship builders to better understand and manage the Navy's industrial base, it collected data on nearly 10,000 Tier 2 suppliers. To reduce this total to a manageable number, NAVSEA conducted extensive research to identify and then focus efforts on the top 850 suppliers of concern.[76] These suppliers were assessed along on a variety of factors, such as the size of the company, its financial health, workforce considerations, and capacity.

Third-Party Providers Fill in the Information Gap

Currently, the vacuum of supplier data is being filled by third-party providers, such as FactSet, S&P Capital IQ, and Bloomberg.[77] Through a variety of proprietary methods, the third-party providers make connections between prime contractors and likely Tier 2 companies. The providers iterate this process down the supply chain, building out likely firm relationships at each level that are one layer up and one layer down. These connections are made using a variety of sources: company press releases, SEC Form 10-K filings, annual reports, bills of lading in customs forms, etc. Financial analysts are the primary audience for the capabilities from the third-party providers, and fees for annual licenses to use these data sets can vary.[78] However, as supply chain concerns grow in national security, the providers have extended their reach to DoD customers. These third-party sources should be used with caution, because, depending on the

---

[74] Frank Camm, Thomas C. Whitmore, Guy Weichenberg, Sheng Tao Li, Phillip Carter, Brian Dougherty, Kevin Nalette, Angelena Bohman, and Melissa Shostak, *Data Rights Relevant to Weapon Systems in Air Force Special Operations Command*, Santa Monica, Calif.: RAND Corporation, RR-4298-AF, forthcoming.

[75] Interview with Air Force Nuclear Weapons Center (AFNWC) personnel, January 28, 2019.

[76] Interview with NAVSEA personnel, May 31, 2019.

[77] FactSet website, undated; S&P Global Market Intelligence website, undated; Bloomberg Professional Services website, undated.

[78] Wall Street Prep, Inc., "Bloomberg vs. Capital IQ vs. FactSet vs. Thomson Reuters Eikon," webpage, undated. The Securities Exchange Act of 1934 requires companies to file an SEC Form 10-K if they are listed on a U.S. exchange, have greater than $10 million in assets, and have at least 2,000 owners or 500 owners who are not accredited investors (Pub. L. 73-291, Securities Exchange Act of 1934, June 6, 1934). Most, but not all, companies that file a 10-K are publicly owned. See SEC, "Exchange Act Reporting and Registration," webpage, October 24, 2018.

methodology, found connections can be old and therefore inaccurate or may not be relevant for particular weapon systems. Also, third-party assessments derived from this data, such as risk scores, might not be tailored to assess DoD- or program-specific risks, which might vary from the risks with which the providers' traditional commercial audiences might be most concerned.

Recommendation: Collect Supply Chain Data from Contractors when Appropriate

Program offices may benefit from routinely collecting the BOM or IPB from contractors. Initially limiting such a policy to only programs identified as meriting additional SCRM attention could help with assessing the costs and benefits of introducing such a requirement more broadly. The determination on whether to collect data for all major defense acquisition programs can be evaluated after understanding the cost for collecting such data and determining whether the insights they provide are worthwhile.

For cases in which the USAF determines it is worthwhile to collect this data, the contract language should not only require the BOM but also include collecting the names of suppliers that provide the parts on the BOM. These supplier data would allow the government to gain insights into both the prime (Tier 1) contractor, and the first layer of subcontractors (Tier 2). For subcontracts that exceed a certain threshold, USAF could request that large subcontractors deliver BOMs as well, thereby gaining insight into some Tier 3 suppliers. Because BOMs and suppliers continue to evolve, the contract would require periodic updates (e.g., annually) to ensure the most up-to-date information.

Additionally, the contract can require a listing of items that can critically affect the reliability of contract end items—also known as a critical items list—to help program offices prioritize parts on the BOM.

6. Data on Suppliers Are Disaggregated and Difficult to Analyze

This chapter identifies a broad range of data sets of potential utility to USAF SCRM but highlights the challenges government personnel might have in accessing and analyzing them to assess supply chain risks.

Among the challenges, even when the suppliers for a weapon system are actually known, only a limited amount of information may be readily available to help government personnel responsible for SCRM conduct risk assessments on the companies.

Given these challenges, it is demanding a lot of SCRM analysts to chase down these data. To the extent that individual program offices are employing SCRM, there is a likely a lot of duplication of effort. As a result, we recommend that USAF consider developing a comprehensive data management plan to guide the collection and integration of the variety of data sources to help SCRM analysts more effectively execute their tasks. Ideally, this should be a whole-of-government effort, given that the data reside in a variety of agencies and commissions. However, in the absence of a larger effort, USAF can begin the process to empower their acquisition professionals.

Some Supply Chain Risk Management–Relevant Data Are Readily Available for Public Companies

Federal securities laws require public companies to disclose information on an ongoing basis. For example, domestic companies must submit annual reports on Form 10-K, which provides a comprehensive overview of the company's business and financial condition and includes audited financial statements.[79] These reports can also provide information on joint ventures and whether the company has subsidiaries based in other countries. The SEC collects these reports in their Electronic Data Gathering, Analysis, and Retrieval (EDGAR) system.[80] However, many private businesses are exempt from this reporting requirement, and such initiatives as DoD's small business contracting goals mean that very little information is publicly available about many of the businesses with which the government contracts.[81]

To approximate what fraction of DoD spending is associated with companies that submit 10-K reports, we downloaded all 2018 10-K reports and compared business names with those listed in FPDS. Using the USA Spending data portal, we downloaded all DoD transactions in

---

[79] SEC, "Form 10-K," webpage, June 26, 2009.

[80] SEC, "EDGAR: Company Filings." webpage, undated.

[81] U.S. Department of Defense, Office of Small Business Programs, "Small Business Program Goals and Performance," webpage, 2020.

FY 2018.[82] Our DoD spending data set contains a total of 47,686 unique company names. It is not possible to join EDGAR and USA Spending data through a common key in a simple one-to-one fashion because the data sources use different conventions to identify companies. Instead of trying to match companies using identifier codes, our solution was to match the company names directly. However, the company names are not quite the same between data sets either. For example, Boeing may be called "BOEING CO" in EDGAR and "THE BOEING COMPANY" in USA Spending data. Given the size of the data set, we could not individually review each company name, but we compared company names in an automated fashion using the lexical technique of approximate string matching (ASM; also known as *fuzzy matching*). The analysis indicates that about 45 percent of DoD spending on prime contracts is with companies that filed 10-Ks in 2018. The appendix has a detailed description of this analysis.

Interestingly, the number of unique companies filing 10-K reports has steadily decreased since 2009. The black line in Figure 6.1 indicates that total filings decreased about 30 percent between 2009 and 2018. The green bars indicate new companies that filed, and the red bars indicates companies that filed in one year but not the next. It is unclear if the trend downward is the result of public companies becoming private or the result of mergers and acquisitions.

Figure 6.1. Number of SEC Form 10-K Filings, by Year

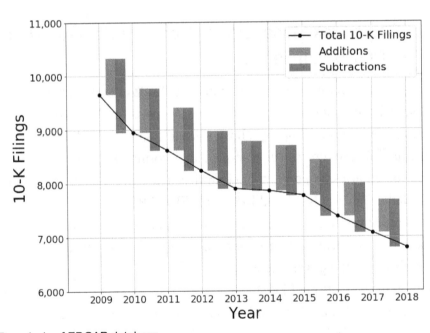

SOURCE: RAND analysis of EDGAR database.

Although the 10-K filings have consistent requirements, which are detailed in Code of Federal Regulations, Title 17, Section § 229.101, the actual implementation of the requirements

---

[82] This included all transactions that designated DoD as the awarding agency (coded as "9700"). See USA Spending, "Custom Award Data," webpage, undated.

can vary.[83] For example, Rockwell Collins—a company that designs and produces communications and aviation systems for military customers—lists joint ventures under the section entitled "Part I, Item 1. Business." Of note for DoD, four of the company's eight joint ventures are with Chinese entities:[84]

- ACCEL (Tianjin) Flight Simulation Co., Ltd, a joint venture with Beijing Bluesky Aviation Technology, a subsidiary of the Aviation Industry Corporation of China (AVIC), to develop and build commercial flight simulators in China
- ADARI Aviation Technology Limited, a joint venture with Aviation Data Communication Corporation Co., LTD, operates remote ground stations around China and develops certain content delivery management software
- AVIC Leihua Rockwell Collins Avionics Company, a joint venture with China Leihua Electronic Technology Research Institute, a subsidiary of the AVIC, which provides integrated surveillance system products for the C919 aircraft in China
- Rockwell Collins CETC Avionics Co., Ltd., a joint venture with CETC Avionics Co., Ltd. to develop and deliver products for the C919 program.

Like Rockwell Collins, Honeywell—a supplier of products, software, and services for military aircraft—also has joint ventures with Chinese entities. Honeywell recently announced a joint venture with FLUX, a provider of warehouse management and related supply chain software in China.[85] Additionally, HonFei Flight Technology Co., Ltd., is a joint venture between Honeywell and AVIC Flight Automatic Control Research Institute (FACRI) that develops and produces flight control systems for commercial aircraft, including for the C919 program.[86] Neither of these joint ventures is mentioned in Honeywell's 2018 SEC Form 10-K report.

Although there might be SCRM value to maintaining awareness of foreign business relationships, the information provided in 10-K reports is not consistently reported in a manner that provides insights into all companies that file reports. Similarly, some, although not all, businesses indicate in their 10-K report whether their manufacturing facilities are located abroad, but, without a detailed analysis, it is unclear whether such information is relevant to a particular weapon system.

Effective September 2018, the SEC expanded the definition of "smaller reporting company," which is intended "to reduce compliance costs for these registrants and promote capital

---

[83] Code of Federal Regulations, Title 17, Commodity and Securities Exchanges, Section 229.101, Description of Business.

[84] Rockwell Collins, Inc., SEC Form 10-K for the Fiscal Year Ended September 30, 2018, Cedar Rapids, Ia., November 26, 2018.

[85] Mark Macaluso, "Honeywell to Invest in Leading Chinese Supply Chain Software Provider and Form a New Joint Venture to Pursue Opportunities Outside China," press release, Honeywell website, December 7, 2017.

[86] Renata Gao, "Honeywell and AVIC FACRI Hold Opening Ceremony for Honfei's New Plant," press release, Honeywell website, March 25, 2017.

formation."[87] The expanded definition means that more SEC registrants would qualify to be exempt from submitting a "less extensive narrative disclosure," which, in turn, also implies the limitations of relying on 10-K filings for a comprehensive picture of a business.[88]

The 10-K reports are notable in that they provide detailed financial information, which would be valuable for understanding the financial health of a company and whether it might be in jeopardy of bankruptcy or a candidate for mergers and acquisition; however, many government analysts are not provided guidance on how to read and interpret that data.

Some Supply Chain Risk Management–Relevant Data Are Located Throughout Government

ICD 731 established and defined the SCRM requirements for mission-critical products, materials, and services used in the intelligence community.[89] The directive requires a risk assessment for acquisitions that the heads of the intelligence community elements deem to be mission critical. A risk assessment consists of a threat assessment of the "proposed contractor, sub-contractor, or vendor (including identified sub-vendors)"; a vulnerability assessment; and an assessment of the potential adverse effects. In turn, Appendix A of ICS 731-02 established minimum information requirements for preparation of a supply chain threat assessment.[90] The information worksheet for conducting the assessment includes collecting data on the company's ownership structure, indications of foreign investment or business ventures with foreign influence, and whether the company has a history of compliance with the various U.S. rules and regulations of doing business. Notably, the worksheet asks whether the company has been compliant with the International Traffic in Arms Regulations (ITAR), which governs the control of exporting defense and military-related technologies outside the United States,[91] and the Foreign Corrupt Practices Act (FCPA), which makes bribing foreign government officials to assist in obtaining or retaining business illegal.[92]

---

[87] SEC, "Smaller Reporting Company Definition," Washington, D.C. June 28, 2018.

[88] SEC, "Smaller Reporting Companies," webpage, July 24, 2019.

[89] ICD 731, 2013.

[90] ICS 731-02, 2016.

[91] Export controls are governed by four different legislative authorities. Dual-use technologies are covered under the Export Controls Act of 2018 (Subtitle B, Part 1 of Pub. L. 115-232, 2018) and the International Emergency Economic Powers Act of 1977 (Title II of With Respect to the Powers of the President in Time of War or National Emergency, October 28, 1977). Munitions are covered by Arms Export Control Act of 1968, 1976 (Title II of Pub. L. 94–329, International Security Assistance and Arms Export Control Act of 1976), and nuclear capabilities are covered by the Atomic Energy Act of 1954 (Pub. L. 83-703, Atomic Energy Act of 1954, August 30, 1954). Ian F. Fergusson and Paul K. Kerr, *The U.S. Export Control System and the Export Control Reform Initiative*, Washington, D.C.: Congressional Research Service, R41916, April 5, 2019.

[92] Pub. L. 95-213, Foreign Corrupt Practices Act of 1977, December 19, 1977.

The intelligence community has identified companies that have violated the rules associated with ITAR and FCPA as carrying a higher level of risk and relevant to SCRM. Although violations of ITAR or FCPA should not in and of themselves indicate a problematic supplier, as the circumstances of those violations are relevant, but this directive requires an IC SCRM analyst to collect data on whether a company has had any ITAR or FCPA violations to assess potential risk.

To collect the data, an IC SCRM analyst will need to collect information from four different agencies. Two agencies track and report ITAR violations—the Department of State's Directorate of Defense Trade Controls and the Department of Commerce's Bureau of Industry and Security. Similarly, two agencies track and report FCPA violations—the Department of Justice (DOJ) and the SEC.

The Directorate of Defense Trade Controls provides a website to search the ITAR violations the directorate tracks, but the website reports business names, rather than Commercial and Government Entity (CAGE) codes,[93] so violators must be matched manually because of variations in naming conventions.[94] The Bureau of Industry and Security reports its violations in annual reports to Congress.[95] As a result, intelligence community SCRM analysts assessing a supplier's risk must download each annual report individually and search the reports for the supplier(s) that are being assessed. Again, business names, rather than unique CAGE codes, must be matched manually.

DOJ is responsible for FCPA criminal enforcement, and SEC for civil enforcement. Both agencies list violators on their websites. DOJ lists violation cases under company or individual last name or by the year the case was filed.[96] SEC lists enforcement actions by date.[97] Some cases have both criminal and civil charges filed and can show up on both the DOJ and SEC lists. As in the case of the export violations, business names, rather than CAGE codes, are reported, so violators must be matched manually because of variation in naming conventions.

For example, Alcoa Inc. does business with DoD and is registered with a Data Universal Numbering System (DUNS) number.[98] In 2014, Alcoa Inc. agreed to settle both SEC charges and DOJ charges for foreign bribery, and the violations are listed as Alcoa and Alcoa World

---

[93] The CAGE code is a five-digit alphanumeric identification number used within the federal government, assigned by the Department of Defense's DLA. The CAGE code provides a standardized method of identifying a given legal entity at a specific location.

[94] U.S. Department of State, "Penalties & Oversight Agreements," webpage, 2020.

[95] U.S. Department of Commerce, Bureau of Industry and Security, annual report archive, 2009.

[96] DOJ, "Enforcement Actions," webpage, July 2, 2020.

[97] SEC, "SEC Enforcement Actions: FCPA Cases," webpage, May 13, 2019.

[98] DUNS number is a nine-digit unique identification code that is similar to a CAGE code. DLA assigns CAGE codes per legal entity at an individual physical address (i.e., the same entity at the same physical address will not be assigned two or more CAGE codes), while a DUNS number is assigned to the lowest organizational level with a unique, separate, and distinct operation.

Alumina on the SEC and DOJ websites, respectively. In this case, it is relatively easy to make connections, but the example illustrates how relying on naming conventions can quickly become problematic.

Under current direction from the intelligence community, a SCRM analyst must do extensive research to answer two relatively simple questions. It is unclear how frequently these data are being accessed, but one could imagine a relatively streamlined database in which an analyst could input a CAGE code and quickly determine whether violations have occurred, thereby greatly reducing the workload burden on SCRM analysts. Nothing precludes such a database from being developed; we were able to create a nascent version for FCPA violations. Such an effort would require an organization within DoD to create and maintain it, although other agencies, such as the Department of State, might also benefit from access.[99]

Some Supply Chain Risk Management–Relevant Data Have Limited Distribution

Returning to the guidance set forth in ICS 731-02, much of the information sought in a supply chain threat assessment of a potential company with which a mission-critical system might contract for goods and services is an understanding of its foreign ownership or influence. The expectation being a company that has significant foreign investment or influence might be targeted by foreign intelligence entities or other adversaries to compromise or exploit the IC supply chain.

Committee on Foreign Investment in the United States

One mechanism with which the U.S. government tracks foreign investment in U.S. businesses is CFIUS. CFIUS was established by executive order in 1975 for federal oversight of foreign investment in the United States as a matter of national security.[100] Through interagency reviews of *covered transactions*,[101] CFIUS provides means to identify and mitigate risks to national security and allows the President to block transactions that pose a credible threat to U.S.

---

[99] Steve Lahr, "State Department Yields on PC's from China," *New York Times*, May 23, 2006.

[100] Executive Order 11858, *Foreign Investment in the United States*, Washington, D.C.: The White House, May 7, 1975.

[101] A covered foreign investment transaction is defined as any merger, acquisition, or takeover "that could result in foreign *control* of any United States business, including such a merger, acquisition, or takeover carried out through a joint venture" that might threaten or impair national security [emphasis added] (James K. Jackson, *The Committee on Foreign Investment in the United States (CFIUS)*, Washington, D.C.: Congressional Research Service, RL33388, August 6, 2019.) National security considerations include issues relating to homeland security, such as critical infrastructure and critical technologies applications. Additionally, the President may consider industries that more broadly affect the capability and capacity of the United States to meet the requirements of national security and whether "a covered transaction is likely to expose, either directly or indirectly, personally identifiable information, genetic information, or other sensitive data of United States citizens to access by a foreign government or foreign person that may exploit that information in a manner that threatens national security" (Jackson, 2019).

national security.[102] USAF may be called on for input to a review, and the office of the Deputy Assistant Secretary of the Air Force for Acquisition Integration (SAF/AQX) manages that process as of this writing. USAF's participation in the review process conveys both a responsibility and an opportunity through which USAF can, within statutory and practical limits, gain greater insight into risks in its supply chain. Arguably, participation in the CFIUS review process constitutes a dimension of SCRM, but data on transactions, including the identities of the parties to a transaction, have strict requirements for confidentiality and use.[103]

Recent legislation on foreign investment in the United States, FIRRMA, both codifies past practices and introduces substantial changes to investment reviews under CFIUS. FIRRMA has, for example, introduced mandatory reporting for some transactions—previously, all reporting was voluntary—and extended CFIUS authorities to include *noncontrolling* investments in U.S. businesses involved in critical technology, critical infrastructure, or collecting sensitive data on U.S. citizens.

Under FIRRMA, USAF will likely have greater responsibility, inasmuch as CFIUS will reach farther and cover more transactions, and greater opportunity. Speaking to responsibilities, USAF can engage more productively in the CFIUS review process the more USAF understands about the composition of its supply chain across sectors. If SAF/AQX or the organizations that SAF/AQX tasks for input lacks knowledge of whether or how the parties to a transaction factor into its supply chain, it cannot effectively assess the potential consequences, including risks, of a transaction. Regarding opportunities, the committee does not review all inbound foreign investment, but the process stands to capture detailed information on transactions of interest to USAF that could shape its understanding of its supply chain and the risks therein, which could, in turn, better position USAF to assess future transactions.

CFIUS data may be relevant for SCRM, but given the highly sensitive nature of these data and the explicit provisions for confidentiality and use, USAF must consult with legal staff on the appropriateness of any access, handling, or application. Currently, these restrictions prevent DoD from widely disseminating information on the companies that have been reviewed as part of CFIUS, so a SCRM analyst may not be aware that a potential supplier has foreign investors aside from what might be publicly reported in the news and trade magazines.

Intelligence and Counterintelligence Products

Intelligence products are important for SCRM because these reports can help program offices assess the level of the intent an adversary might have to disrupt the program's supply chain. DIA

---

[102] For a discussion of the CFIUS process, including recent changes, see Jackson, 2019.

[103] See, e.g., U.S. Code, Title 50, War and National Defense, Section 4565, Authority to Review Certain Mergers, Acquisitions, and Takeovers; U.S. Code, Title 50, War and National Defense, Appendix, Section 2170, Authority to Review Certain Mergers, Acquisitions, and Takeovers, Subsection (g) Additional Information to Congress; Confidentiality.

and the service intelligence organizations produce intelligence threat assessments that support major defense acquisition programs.[104] However, intelligence producers have several structural limitations concerning analyses looking at threats to and through the supply chain, especially when U.S. persons or entities are involved. Foreign intelligence analysts focus on threats to the United States from foreign adversaries and entities; these analysts cannot collect and produce intelligence on U.S. individuals or entities.[105] This restriction includes U.S. companies that have facilities in a foreign country. As a result, they do not have the authority or the information to respond to questions concerning these specific threats.

Counterintelligence activities, which are aimed at thwarting our adversaries' spies, are important for SCRM because these activities can help identify suppliers that might be targeted by foreign intelligence entities for one reason or another. DIA also conducts counterintelligence activities to support SCRM in research, development, and acquisition.[106] Counterintelligence analysts have greater accesses and authorities in collecting information on U.S. businesses for these purposes, but these too are limited. In this case, the service counterintelligence and security entities (Air Force Office of Special Investigations [AFOSI)], for the Air Force) would be the source of this analysis.

Our review of intelligence and counterintelligence products that support SCRM practices revealed the disaggregate dissemination practices of this type of information. Because the DoD supply chain involves U.S. persons and entities, Intelligence Oversight procedures apply.[107] These procedures govern the activities of intelligence and counterintelligence organizations pertaining to U.S. persons' information.[108] For this reason, access to some products may be limited and available by request only or may be carefully written to allow wide dissemination. Intelligence oversight further prohibits blacklisting companies on the basis of intelligence information. As an example of complying with intelligence oversight while still providing valuable information for SCRM analysts, AFOSI baseline threat assessments provide insights on

---

[104] DIA Instruction 5000.002, *Intelligence Threat Support for Major Defense Acquisition Programs*, Defense Intelligence Agency, June 19, 2019.

[105] DoDM 5240.01, *Procedures Governing the Conduct of DoD Intelligence Activities*, Washington, D.C.: Office of the Deputy Chief Management Officer of the Department of Defense, August 8, 2016.

[106] DoDI, O-5240.24, *Counterintelligence Activities Supporting Research, Development, and Acquisition*, Washington, D.C.: Under Secretary of Defense for Intelligence, June 8, 2011, change 1, October 15, 2013, Not available to the general public.

[107] DoDM 5240.01, 2016; DoD Directive 5148.13, *Intelligence Oversight*, Washington, D.C.: Office of the Deputy Chief Management Officer of the Department of Defense, April 26, 2017; and Executive Order 12333, *United States Intelligence Activities*, Washington, D.C.: The White House, December 4, 1981.

[108] Executive Order 12333, 1981; Executive Order 12334, *President's Intelligence Oversight Board*, Washington, D.C.: The White House, December 4, 1981; DoD Directive 5148.13, 2017. These procedures enable DoD to conduct authorized intelligence activities in a manner that is legal and focus on intelligence activities that include collection, retention, and dissemination of U.S. Persons Information and intelligence methods that include electronic surveillance, concealed monitoring, physical searches, searches of mail and use of mail covers, physical surveillance, and undisclosed participation in organizations.

potentially corruptible products, business practices, or external influences for microelectronics without addressing specific companies.[109]

Additionally, intelligence and counterintelligence organizations primarily operate in the Joint Worldwide Intelligence Communications System (JWICS) network environment, while the rest of DoD, including acquisitions and nonintelligence organizations, primarily use the Secret Internet Protocol Router Network (SIPRNet) for communication and information-sharing. Although many intelligence products are located on SIPRNet, more intelligence analytical products reside on JWICS and may be available on request. Acquisition professionals requiring access will need to have the appropriate clearance and access levels to be able to request this information or to view it on their own on JWICS.[110]

Some Supply Chain Risk Management–Relevant Data Are in Formats Not Conducive to Analysis

As in the case of ITAR and FCPA violations, SCRM-relevant data can be available but in a manner that requires substantial effort for an analyst to collect and assess, whether it be individual reports or manually matching business entities. This issue is not unique, and we wish to highlight the extent to which information that might be useful for SCRM is collected in a manner that creates barriers for its use in SCRM.

Counterfeit parts in the military's supply chain have been and continue to be a key concern for DoD, and evaluating the risk associated with counterfeit parts is an important component of SCRM.[111] Counterfeit parts are problematic in two ways. First, a counterfeit part can be lower-quality products, which may affect system reliability. Second, a counterfeit part could include a nonattributable backdoor foreign intelligence entities.

Per Section 818 of the FY 2012 NDAA, DoD is required to have its personnel and contractors report suspected counterfeit electronic parts to a partnership between government and industry for sharing technical information, the Government-Industry Data Exchange Program (GIDEP).[112] DoD has codified this requirement in both DFARS and DoDI 4140.01.[113] We were unable to assess compliance with DoD policies and regulations, but GAO has expressed concern about inadequate oversight.[114] GIDEP includes reports on parts failure, engineering reports, and

---

[109] Interview at AFOSI, February 5, 2019.

[110] Interviews with AFMC, November 7, 2018; AFOSI, February 5, 2019; SCRM TAC, February 28, 2019; and 448th Supply Chain Management Wing, June 19, 2019.

[111] GAO, *Counterfeit Parts: DOD Needs to Improve Reporting and Oversight to Reduce Supply Chain Risk*, Washington, D.C., GAO-16-236, February 2016.

[112] Pub. L. 112–81, National Defense Authorization Act for Fiscal Year 2012, December 31, 2011.

[113] Code of Federal Regulations, Title 48, Federal Acquisition Regulations System, Section 252.246-7007, Contractor Counterfeit Electronic Part Detection and Avoidance System; DoDI 4140.01, 2019.

[114] GAO, 2016.

suspect counterfeit reports. The suspect counterfeit reports include document title, number, date, part number, National Stock Number (NSN) of the suspect part,[115] manufacturer, CAGE, supplier, and supplier CAGE code, but the completeness of these sections is variable. These reports are provided in portable document format (PDF) files that must be individually downloaded. Figure 6.2 shows an example submission.

Per the GAO report, several DoD officials envisioned GIDEP as an early warning system to prevent counterfeit parts from entering the defense supply chain.[116] GAO noted that reports on suspect counterfeit parts were not made available to industry in a timely and comprehensive manner. We also noted a secondary hurdle associated with the way in which the data is collected: The individual reports prevented analysts from easily combining the data to see trends.

If the data in GIDEP reports could be combined in a database, SCRM analysts could potentially identify parts that are more frequently counterfeited and, therefore, require a more robust mitigation strategy. Additionally, data from GIDEP could note problematic suppliers that might merit more attention for oversight. Both types of analyses could help fulfill the program's intent as an early warning system; however, for these analyses to occur, each PDF must be downloaded, the data scrapped, and a database created. We were able to create a prototype database, but the process was nontrivial and likely outside the capabilities we would expect a standard SCRM analyst to have.

As in the case of ITAR and FCPA violations, we would envision that an organization within DoD would need to create and maintain a database of suspected counterfeit parts so that SCRM analysts could ask prime contractors (or review the BOM) to determine whether their systems had parts that are more frequently counterfeited or if the prime contractor used a problematic supplier. Additionally, despite DoD's direction, information regarding suspected counterfeit parts was collected in a variety of locations, such as the Navy's Product Data Reporting and Evaluation Program and ERAI, a company that monitors, investigates, and reports issues affecting the global electronics supply chain.[117] The GAO report noted that ERAI "had significantly more suspect counterfeit reports than GIDEP."[118]

---

[115] The NSN is a 13-digit identification number, the first four of which constitute the Federal Supply Class Group (FSCG) and the last nine of which make up the NIIN, a unique code used to distinguish each part.

[116] GAO, 2016.

[117] Department of the Navy, Product Data Reporting and Evaluation Program (PDREP), "Who We Are," webpage, undated; ERAI, Inc., homepage, undated.

[118] GAO, *Counterfeit Parts: DOD Needs to Improve Reporting and Oversight to Reduce Supply Chain Risk*, Washington, D.C., GAO-16-236, February 2016.

Figure 6.2. Example Government-Industry Data Exchange Program Report, Redacted

Distribution is not authorized outside of the GIDEP participant's organization.

GOVERNMENT - INDUSTRY DATA EXCHANGE PROGRAM	
ALERT	

1. TITLE (Class, Function, Type, etc.)	2. DOCUMENT NUMBER
▨	B9B-A-18-04
	3. DATE (DD-MMM-YY)
	11-JUL-18

4. MANUFACTURER AND ADDRESS	5. PART NUMBER	6. NATIONAL STOCK NUMBER
▨	▨	Not Available
	7. SPECIFICATION	8. GOVERNMENT PART NUMBER
	Not Available	Not Available
	9. LOT DATE CODE START	10. LOT DATE CODE END
	▨	▨

11. MANUFACTURER'S POINT OF CONTACT	12. CAGE	13. MANUFACTURER'S FAX
Not Applicable	Not Available	Not Applicable
14. MFR. POC PHONE	15. MANUFACTURER'S E-MAIL	
Not Applicable	Not Applicable	

16. SUPPLIER	17. SUPPLIER ADDRESS	18. SUPPLIER CAGE
▨	▨	Not Available

19. PROBLEM DESCRIPTION / DISCUSSION / EFFECT

▨

Note: *The manufacturer identified in block 4 is the entity whose product may have been counterfeited. This reporting convention is necessary to facilitate GIDEP database searches for suspect counterfeit products and is by no means intended to imply that the manufacturer identified in block 4 is involved with the suspect product.*

20. ACTION TAKEN/PLANNED

▨

21. DATE MFR. NOTIFIED/ SUPPLIER NOTIFIED	22. MFR./SUPPLIER RESPONSE	23. ORIGINATOR ADDRESS/POINT OF CONTACT
15-JUN-18	[X] REPLY ATTACHED [ ] NO REPLY	▨

24. GIDEP REPRESENTATIVE	25. SIGNATURE	26. DATE
▨	Signature on File	11-JUL-18

GIDEP Form 97-1 (September 2009)

Please refer to the complete distribution policy at the GIDEP member's website.

SOURCE: GIDEP.

Recommendation: Comprehensive Data Management Strategy

If USAF wishes to implement SCRM for its new and legacy weapon systems, it would likely be both cost prohibitive and inefficient to expect individual SCRM analysts to expend the effort necessary to collect the relevant data from disparate sources or to develop the tools and databases themselves to execute their tasks effectively. It would behoove USAF to develop a comprehensive data management plan that would guide the collection and integration of the variety of data sources that are scattered throughout government and the public sector. This endeavor could be coupled with a centralized collection of BOM data, thereby providing SCRM analysts a centralized resource to enable timely decisionmaking informed by complete information.

As an example of the types of decisions this resource could inform, consider a CFIUS review. If, as a part of CFIUS, government officials request that the company being purchased provide a list of current customers, SAF/AQX could then reference the centralized database of BOM and associated suppliers to determine whether the company or its customers provide parts to USAF weapon systems. This would enable SAF/AQX to quickly determine whether a company under CFIUS review is in USAF's supply chain down to Tier 3 and potentially Tier 4, which provides much greater insight than they currently have. This knowledge then allows implementation of a mitigation strategy and reduces any risk associated with the merger or acquisition.

Centralized suspected counterfeit reporting combined with BOM data is another example of how a centralized data collection could provide timely information to SCRM analysts in program offices. If data on suspected counterfeit parts from GIDEP, the Product Data Reporting and Evaluation Program, and ERAI were collected in a systematic fashion and merged, SCRM analysts could review a single feed of information on recent counterfeit reports and search their weapon system BOMs for whether any contains the same NSN. This then enables SCRM analysts to proactively respond to potential problems rather than reactively responding to issues as they arise.

It is important to note that a comprehensive data strategy, in and of itself, is not sufficient for understanding supply chain risk, but it is an important initial step in providing SCRM analysts the tools they need to effectively and efficiently execute their roles.

7. Summary and Conclusions

The goals of SCRM are to avoid supply disruptions and, if they occur, to mitigate their effects on USAF. A proactive enterprisewide SCRM approach could potentially help USAF anticipate supply chain risks and develop and implement appropriate mitigation strategies. This report offers recommendations on organization, policy, training, data sources, and data management to support more-effective SCRM.

Below are various recommendations that USAF—and DoD more broadly—may consider if it wishes to develop a robust SCRM capability. As mentioned in Chapter 1, this analysis did not assess the cost-effectiveness of these recommendations, and USAF will need to determine whether the potential benefits of implementing the recommendations are an effective use of limited resources. We based the recommendations on two primary sources: (1) current practice within DoD or industry and (2) interviews with SCRM practitioners and stakeholders across DoD. Additionally, successfully implementing these recommendations requires appropriate expertise that has not yet been identified or cultivated.

Organize to Facilitate Supply Chain Risk Management

Within USAF, policy and guidance that define SCRM roles and responsibilities are dispersed and focus on only a subset of SCRM, such as counterfeits or cybersecurity. As of this writing, part from the SCRM WG—which was established only in 2018—no single organization at the DoD or Air Force level is responsible for directing an enterprisewide SCRM vision. Perhaps as a result, no enterprise-level SCRM policy exists at either organizational level.

USAF can potentially draw on the commercial sector for lessons about how to organize effectively to support SCRM; however, not all industry best practices are applicable to DoD because of differences in attack type, dynamism of demand, and sustainment, to name a few. One organizational construct for SCRM in the commercial sector is to have two levels of management: an executive level and an execution level. Where this construct exists, the executive level is usually in the form of a panel or council composed of key stakeholders from across the company. The execution level may be centralized, with a dedicated SCRM team, or decentralized.

Industry often relies on several mechanisms to manage supply chain risk, many of which rely on information-sharing among firms and their suppliers. By contrast, lack of information-sharing within DoD has often been identified as a challenge for implementing SCRM. When information is not shared, best practices are not disseminated, and efforts may be duplicative.

Following commercial practice, **USAF should consider forming an executive council of SCRM stakeholders to establish policy, set standards, and facilitate information-sharing.**

41

These stakeholders are not necessarily SCRM experts, so **USAF could create an organization with SCRM experts and analysts to support the executive SCRM council.** Such an organization could conduct centralized analyses that are beyond the scope of program offices. This executive council could help establish policy to identify which program merit the additional attention to SCRM.

To facilitate execution of SCRM within the programs identified, **program offices could create supply chain working groups composed of program office personnel, DLA representatives, supply chain management wing personnel, and any other relevant stakeholders.**

When Appropriate, Incentivize and Enforce Supply Chain Risk Management

USAF suppliers are likely to optimize their operations for cost, schedule, and performance, with performance being narrowly defined as reaching certain capability thresholds rather than avoiding supply chain disruptions. It would be a mistake for USAF to assume that its suppliers actively manage their supply chains.

Current policy and guidance are of limited utility in helping USAF incentivize suppliers to engage in risk management and to enforce their participation. For example, 10 U.S.C. 2305 *allows* an agency to consider supply chain risk as a factor in contracting but does not *require* consideration of this risk. DFARS provides clauses that direct suppliers to mitigate supply chain risk, but these clauses do not include an enforcement mechanism.

The Army and the intelligence community both have policies in place to identify programs that merit further scrutiny for SCRM. **USAF may wish to mandate consideration of SCRM and supply chain security as factors in source selection for programs identified as sufficiently important to merit additional attention for SCRM.** This prioritization of SCRM early in source selection may help incentivize suppliers to help manage supply chain risk and enable USAF to monitor those efforts. Third-party auditors or the government will need to objectively ensure that the contractor is fulfilling its obligations to achieve a successful SCRM program.

Provide More Supply Chain Risk Management Training

SMEs emphasized that, as modern weapon systems become increasingly complex, so do their supply chains, and SCRM in general becomes more challenging. In particular, supply chain risk assessments require extensive information. The USAF SCRM workforce needs guidance and training to understand how supply chain risks are evolving and what information to gather to conduct risk assessments and develop mitigations.

As of this writing, the acquisition workforce is not informed about the full range of supply chain risks and how to collect sufficient information to understand them. Several organizations that conduct risk assessments expressed the need for program offices to be better informed and trained on information that should be requested from contractors prior to conducting risk assessments. The Navy currently has training available. **The USAF acquisition workforce may benefit from attending training developed elsewhere within DoD, or USAF may consider developing an ongoing, formal, enterprise-level training curriculum to focus on its own workforce and processes.** It is particularly important to educate acquisition professionals on supply chain risk for programs that are identified as meriting additional SCRM attention.

Collect Data on Lower-Tier Suppliers

SCRM analysts have insight into Tier 1 contractors but limited insight into Tier 2 and below with existing data sources. USAF SCRM analysts can obtain information about suppliers and even subcontractors by including data requests in contracts. Naturally, these data requests would come at an additional cost, and USAF, likely through an executive council, would need to assess under what conditions the costs are worthwhile.

When appropriate, USAF should consider requiring program offices to request BOMs—also known as IPBs—from contractors and major subcontractors and a commensurate list of any suppliers for those parts. The Navy routinely collects these data and has found them invaluable for managing its supply chain risk. If USAF were to collect such information, it would need to be verified by government or third-party audits. Such supplier data would allow the government to gain insight into both the prime contractor (Tier 1) but also the first layer of subcontractors (Tier 2). Depending on the weapon system, acquisition professionals may want to consider requesting a software BOM.

Program offices could benefit from requesting critical items lists to help prioritize parts. The critical items list is frequently provided by the contractor and informed by a weapon system's engineers. These data may already be collected in some form, but SCRM analysts should consider using the data to inform prioritization of SCRM efforts. Contracts should require periodic updates (e.g., annually) to ensure that USAF possesses the most up-to-date information.

Our interviews indicated that there was tension between employing SCRM best practices and the requirement to address obsolescence. **Program offices should consider the value of technical data to SCRM in deciding how much to invest in the delivery of technical data during initial contract negotiation.** Such data would potentially allow programs to assess multiple alternative sources of supply if obsolescence becomes an issue. SCRM analysts can potentially identify alternative sources of supply by accessing the Advanced Component Obsolescence Management system, which identifies parts that are common among weapon systems that use the management system.

We caution USAF SCRM analysts against relying on third-party providers for detailed data on supply chains. Depending on the methodology, third-party providers may rely on data sources that are inaccurate or outdated. Additionally, a company's supplier relationships may not be relevant for a given weapon system, hence the importance of collecting BOM data.

Manage Data to Support Supply Chain Risk Management

SCRM-relevant data are collected and reported by diverse government agencies, and these efforts are not coordinated and managed in a manner that is conducive to their integration and analysis. Several SMEs highlighted the difficulties of integrating a variety of databases to better inform SCRM activities. Program offices are often asked to accomplish more tasks with less resources, and it is demanding a lot of program offices to chase down SCRM-relevant data from disparate sources to effectively execute these tasks. When SCRM is employed at program offices, there is likely considerable duplication of effort to get data in an easily analyzable format.

Industry practice routinely collects such information to inform supplier quality ratings. **USAF may want to develop a comprehensive data management plan to guide the collection and integration of the variety of data sources to help SCRM analysts execute their tasks more efficiently.** Ideally, this should be a whole-of-government effort, given that the data reside in a variety of agencies and commissions. In the absence of a larger effort, USAF can begin the process to empower its own acquisition professionals.

Conclusion

There are many changes USAF and DoD could make to better posture themselves for effectively executing SCRM. Existing policies and authorities are not cohesive; roles and responsibilities are dispersed; personnel charged with SCRM activities lack sufficient guidance, training, and incentives; data needed to inform risk assessments and the development of effective mitigations are dispersed and not managed to support SCRM. By accepting and implementing the recommendations outlined in this report, USAF can begin making strides to address these SCRM shortfalls and position itself to anticipate and effectively mitigate new supply chain risks as they emerge.

Appendix. Approximate String Matching of Company Names

DoD has contracts with many companies that are annually required to file a Form 10-K with the SEC. However, determining how much money DoD pays these companies is nontrivial because the SEC and U.S. spending data sets—EDGAR and USA Spending, respectively—do not share common keys. Our workaround solution was to use an approximate string-matching process to quantify the similarity of company names in the two data sets with a high degree of confidence. The total amount of DoD spending on contracts in FY 2018 was $375.0 billion, and we estimate that DoD paid these companies between $164.3 billion (43.82 percent of the total) and $178.9 billion (47.69 percent) in FY 2018, depending on whether the confidence threshold is set at the low end or high end.

Data Sources

The SEC requires many U.S. companies with large asset holdings to disclose financial and other information annually on a standardized form, 10-K. The 10-K includes many details on a company's financial health and is publicly accessible on the SEC's website through the EDGAR database.[119]

DoD has contracts with many companies that file 10-Ks, and it would be beneficial to know approximately how much money goes to these companies on an annual basis. To do this, we first began by collecting the SEC and DoD spending data. Limiting our analysis to a one-year period, we queried and scraped all company names from 10-K filings on EDGAR in 2018. The total number of unique companies that filed a 10-K in 2018 was 6,800. After getting company names from EDGAR, we collected data on DoD spending. Two main sources exist for tracking government spending, FPDS and USA Spending. Either data source would have been sufficient for this analysis, but, in general, USA Spending is considered more comprehensive because it collates data not only from FPDS but also from other sources.[120] Using the USAspending.gov data portal, we downloaded all DoD transactions in FY 2018.[121] The total number of unique company names in our DoD spending data set is 47,686. The total amount of DoD spending on contracts in FY 2018 was $375.0 billion.[122]

---

[119] SEC, "How to Read a 10-K," webpage, July 21, 2011.

[120] "What Is the Difference Between FPDS-NG and USA Spending.gov?" FPDS-NG FAQ, undated.

[121] This meant all transactions which designated DoD as the awarding agency (coded as "9700"). See "Custom Award Data," webpage, undated.

[122] Here, we only consider positive transactions; negative values represent reductions to the original contract obligation but do not represent dollar exchanges.

Marrying these data sets and determining how much of the DoD budget goes to companies that file 10-Ks is a nontrivial, multistep process. Because the data sources use different conventions to identify companies, it is not possible to join EDGAR and USA Spending data through a common key in a simple one-to-one fashion. The SEC assigns a unique ten-digit number, called a Central Index Key, to each company; DoD uses CAGE codes to distinguish companies; and the U.S. government as a whole uses DUNS numbers. CAGE codes and DUNS numbers are tied to company addresses, which means companies with multiple locations can have several different identifiers.[123] Because of the complex and myriad relationships among parent companies and their subsidiaries, there is no established method for linking a company's Central Index Key to all its associated DUNS numbers and CAGE codes.

Matching Methodology

Instead of trying to match companies using identifier codes, our solution was to match the company names directly. However, the company names are also not quite the same between data sets. For example, Boeing may be called "BOEING CO" in EDGAR and "THE BOEING COMPANY" in USA Spending data. Common sense tells us that these are the same companies. But there are thousands of companies in the data set, so it would be impossible to match them by hand. Instead, we can scale this process with a high degree of accuracy by using the lexical technique of ASM. This technique makes it possible to compare many different text samples and quantify the similarities of their patterns using a weighted scoring system.

Before introducing our ASM process, it is important to clarify which company names we chose to use. The USA Spending data dictionary has three different data fields referring to the name of the company receiving the transaction.[124] There is the *recipient name*, the *recipient doing business as (DBA) name*, and the *recipient parent name*. The data dictionary says that the recipient name is the legal name of the awardee filed information with individual states, and the DBA is essentially an alias for the recipient name. The recipient parent name, on the other hand, is the ultimate parent of the awardee and is tied to the company's global DUNS number. Many of these companies are subsidiaries, and we prefer the most general name for the company, so we choose to use the recipient parent name.[125] Table A.1 shows examples of some different names in USA Spending used for award recipients operating under Boeing's auspices.

---

[123] For example, searching for "Boeing" in the Dun & Bradstreet search engine produces hundreds of results just in the state of Washington.

[124] USA Spending, "Data Dictionary," webpage, undated b.

[125] The recipient parent name is not always available, and when it is not, we revert to the recipient name.

**Table A.1. Example of Naming Conventions Used in USA Spending Data
for Companies Associated with Boeing**

Recipient Name	Recipient DBA Name	Recipient Parent Name
ARGON ST. INC.	—	THE BOEING COMPANY
AVIALL SERVICES, INC.	AVIALL	THE BOEING COMPANY
BOEING AEROSPACE OPERATIONS, INC.	BOEING	THE BOEING COMPANY
BOEING SIKORSKY AIRCRAFT SUPPORT	—	THE BOEING COMPANY

The ASM process can differ depending on the data types being matched, but an important first step is to normalize the data. The company names in their original forms have many irregularities or unimportant characters that should be removed. We used the Python programming package called *cleanco* to automatically strip away terms found at the end of an organization name that identify the organization type, such as *LLC*, *Inc.*, and *Corp.*[126] These words are ultimately not important for identifying matches because they have very little intrinsic meaning.[127] Next, we stripped all punctuation from the name to remove other insignificant characters, such as ampersands, commas, and quotation marks. Then, we parsed the names into individual words, which are often referred to as *tokens* in the field of lexical analysis. After tokenizing the names, we made all characters lowercase and removed other *stopwords*, such as *is*, *and*, and *are*—once again, these provide little meaning for string matching.[128]

Having tokenized the names, we could focus on individual words rather than the string as a whole. For many ASM methods, this is when a concept referred to as *edit distance* becomes important. This is the minimum number of operations required to make one text string match another. For example, it would require a one-character substitution to change the name "THE BAEING COMPANY" to "THE BOEING COMPANY." The fewer the edits, the smaller the edit distance. Popular measurements are the Jaro-Winkler distance and the Levenshtein distance, which permit the use of different edit operations, such as insertion, deletion, and substitution.[129] However, for our purposes, these methods were not appropriate because the company names are from official records and are unlikely to have misspellings. The more likely scenario is that some words will be missing or appear in a different order (e.g., "BOEING CO"). Thus, our process disregarded word order but took into account the rarity or overall importance of certain words. Table A.2 shows examples of the results of the tokenization process.

---

[126] "Cleanco 2.0.1," Python Package Index (PyPi) website, undated.

[127] We acknowledge that there is a possibility that two completely separate entities may have the same name with different identifiers (e.g., Acme Inc. and Acme Corp.), but we assessed that the likelihood was relatively small and the effects on our qualified findings even smaller.

[128] Here, we used the Stopwords Corpus from the Natural Language Toolkit ("Natural Language Toolkit," NLTK 3.5 Documentation website, April 13, 2020).

[129] William W. Cohen, Pradeep Ravikumar, and Steven E. Feinberg, "A Comparison of String Distance Metrics for Name-Matching Tasks," paper, American Association for Artificial Intelligence, 2003.

Table A.2. An Example of Company Names as They Appeared Originally and After Tokenization

USA Spending Name	SEC 10-K Name	USA Spending Tokens	SEC 10-K Tokens	Match Score
NORTHROP GRUMMAN CORPORATION	NORTHROP GRUMMAN CORP	['grumman', 'northrop']	['grumman', 'northrop']	1.0000
NORTHROP GRUMMAN SYSTEMS CORPORATION	NORTHROP GRUMMAN CORP	['grumman', 'northrop', 'systems']	['grumman', 'northrop']	0.9834
NORTHROP GRUMMAN DEFENSE MISSION SYSTEMS INC.	NORTHROP GRUMMAN CORP	['defense', 'grumman', 'mission', 'northrop', 'systems']	['grumman', 'northrop']	0.8586

To determine the rarity of words, we used a variation of a well-established term-weighting function from computational linguistics called *term frequency–inverse document frequency*.[130] This calculation essentially weighs the importance of words by taking the number of occurrences of each token in our full list of tokens. Such words as *group* and *services* have relatively high counts, while such words as *Boeing* are quite rare.

Although words with high counts have less overall importance, they should not be overly ignored. Thus, we scaled down large differences in frequency and calculate the *rarity* of a company name as a whole by adding together the inverse square root of each token's frequency (*f*):[131]

$$Rarity(f_1, f_2, \ldots f_n) = \sum_{i=1}^{n} \frac{1}{\sqrt{f_i}}.$$

Summing the rarity of each token in a company name gives its overall rarity. When calculating the similarity of two company names with token sets A and B, we want to calculate the rarity of their common tokens and divide that by the square root of the product of the rarity of all of their tokens:[132]

$$Similarity(A, B) = \frac{Rarity(A \cap B)}{\sqrt{Rarity(A) * Rarity(B)}}.$$

Consequently, sets of tokens that overlap perfectly will receive a similarity score of 1, and those that match less precisely will have lower scores. On visual inspection, the confidence of the match becomes dubious around the 0.85 mark. Recall from Table A.2 that the comparison of Northrop Grumman Defense Mission Systems, Inc., and Northrop Grumman Corp. received a score of 0.86, which is rather low but still probably correct. However, Table A.3 includes examples of scores below 0.85, which may include incorrect matches.

---

[130] For a thorough discussion of this technique, see Stephen Robertson, "Understanding Inverse Document Frequency: On Theoretical Arguments for IDF," *Journal of Documentation*, Vol. 60, No. 5, October 2004.

[131] We used the square root to smooth values in our example, but logarithmic scaling is also very common.

[132] We followed a process very close to the one outlined in the following GitHub walkthrough: "Data Matching Part 3: Match Scoring," *DS lore* blog, July 29, 2016.

Setting the threshold at different levels can establish upper and lower bounds for the amount of money DoD pays to companies that file a 10-K with the SEC. A threshold of 0.80 and above would mean that, at most, $178.9 billion (47.69 percent) of DoD's spending on contracts in FY 2018 went to companies that filed 10-Ks. Setting the highest possible threshold at 1.00 would require perfect token matches and be equal to $164.3 billion (43.82 percent). In our data, the values are not drastically different in this high to low range. This is because the largest prime contracts tend to go to companies that have perfect token matches. Thus, even under the strictest criteria, the differences are minimal, and we can say with high confidence that the amount of DoD money going to these companies could not be much more than one-half of total defense contract spending.

Table A.3. Examples of Scores Below the 0.85 Threshold That May Include Incorrect Matches

USA Spending Company Name	SEC Form 10-K Company Name	USA Spending Tokens	SEC Form 10-K Tokens	Match Score
DATA LINK SOLUTIONS L.L.C.	I LINK INC	['data', 'link', 'solutions']	['link']	0.8053
UNITED EXCEL CORPORATION	Excel Global Inc	['excel', 'united']	['excel', 'global']	0.8094
CHUGACH ALASKA CORPORATION	CHUGACH ELECTRIC ASSOCIATION INC	['alaska', 'chugach']	['association', 'chugach', 'electric']	0.8155
AGILE DEFENSE INC.	AGILE THERAPEUTICS INC	['agile', 'defense']	['agile', 'therapeutics']	0.8085
AEGIS DEFENSE SERVICES LLC	Aegis Holdings Inc	['aegis', 'defense', 'services']	['aegis', 'holdings']	0.8163
HIGH DESERT SUPPORT SERVICES LLC	High Desert Holding Corp	['desert', 'high', 'services', 'support']	['desert', 'high', 'holding']	0.8145
UNIVERSAL CONSULTING SERVICES INC	UNIVERSAL CORP	['consulting', 'services', 'universal']	['universal']	0.8126
POLARIS ALPHA LLC	POLARIS INDUSTRIES INC	['alpha', 'polaris']	['industries', 'polaris']	0.8040
CONSTRUCTION HELICOPTERS INC.	PETROLEUM HELICOPTERS INC	['construction', 'helicopters']	['helicopters', 'petroleum']	0.8112
SPRINT FEDERAL OPERATIONS LLC	SPRINT CORP	['federal', 'operations', 'sprint']	['sprint']	0.8178

Bibliography

Ackoff, Russell L., "From Data to Wisdom," *Journal of Applied Systems Analysis*, Vol. 16, No. 1, 1989, pp. 3–9. As of August 12, 2020:
https://softwarezen.me/wp-content/uploads/2018/01/datawisdom.pdf

AFI—*See* Air Force Instruction.

AFPAM—See Air Force Pamphlet.

AFPD—*See* Air Force Policy Directive.

Air Force Instruction 23-101, *Air Force Materiel Management*, Washington, D.C.: Department of the Air Force, September 9, 2019. As of July 31, 2020:
https://static.e-publishing.af.mil/production/1/af_a4/publication/afi23-101/afi23-101.pdf

Air Force Instruction 18-130, *Cybersecurity Program Management*, Washington, D.C.: Department of the Air Force, February 12, 2020. As of August 6, 2020:
https://static.e-publishing.af.mil/production/1/saf_cn/publication/afi17-130/afi17-130.pdf

Air Force Instruction 63-101/20-101, *Integrated Life Cycle Management*, Washington, D.C.: Department of the Air Force, March 7, 2013, change 2, February 23, 2015 (guidance memorandum dated September 16, 2016). As of August 6, 2020:
https://www.dau.edu/guidebooks/SiteAssets/htmlviewer/supporting%20documents/Acquisition%20Reference%20Desktop/afi63-101_20-101.pdf

Air Force Pamphlet 63-113, *Program Protection Planning for Life Cycle Management*, Washington, D.C.: Department of the Air Force, October 17, 2013.

Air Force Policy Directive (AFPD) 23-1, *Supply Chain Materiel Management*, Washington, D.C.: Department of the Air Force, September 7, 2018. As of July 31, 2020:
https://static.e-publishing.af.mil/production/1/saf_aq/publication/afpd23-1/afpd23-1.pdf

Air Force Supply Chain Risk Management Working Group, "Air Force Supply Chain Risk Management Campaign Plan," undated.

———, "Air Force Supply Chain Risk Management Working Group Charter," April 30, 2019.

Army Regulation 70–77, *Program Protection*, Washington, D.C.: Headquarters, Department of the Army, June 8, 2018. As of July 31, 2020:
https://armypubs.army.mil/epubs/DR_pubs/DR_a/pdf/web/ARN8083_AR70-77_Web_FINAL.pdf

Bacon, David F., Rodric M. Rabbah, and Sunil Shukla, "FPGA Programming for the Masses," *Communications of the ACM*, Vol. 56, No. 4, April 2013, pp. 56–63. As of July 31, 2020: https://researcher.watson.ibm.com/researcher/files/us-bacon/Bacon13FPGA.pdf

Bloomberg Professional Services website, undated. As of July 31, 2020: https://www.bloomberg.com/professional/

Camm, Frank, Thomas C. Whitmore, Guy Weichenberg, Sheng Tao Li, Phillip Carter, Brian Dougherty, Kevin Nalette, Angelena Bohman, and Melissa Shostak, *Data Rights Relevant to Weapon Systems in Air Force Special Operations Command*, Santa Monica, Calif.: RAND Corporation, RR-4298-AF, forthcoming.

"Cleanco 2.0.1," Python Package Index (PyPi) website, undated. As of August 6, 2020: https://pypi.org/project/cleanco/

Code of Federal Regulations, Title 17, Commodity and Securities Exchanges, Section 229.101, Description of Business. As of July 31, 2019: https://www.law.cornell.edu/cfr/text/17/229.101

Code of Federal Regulations, Title 48, Federal Acquisition Regulations System, Section 15.304, Evaluation Factors and Significant Subfactors. As of August 4, 2020: https://www.law.cornell.edu/cfr/text/48/15.304

Code of Federal Regulations, Title 48, Federal Acquisition Regulations System, Section 52.249-14. As of August 6, 2020: https://www.law.cornell.edu/cfr/text/48/52.249-14

Code of Federal Regulations, Title 48, Federal Acquisition Regulations System, Section 52.204-10, Reporting Executive Compensation and First-Tier Subcontract Awards. As of August 6, 2020: https://www.law.cornell.edu/cfr/text/48/52.204-10

Code of Federal Regulations, Title 48, Federal Acquisition Regulations System, Subpart 239.73, Requirements for Information Relating to Supply Chain Risk. As of August 6, 2020: https://www.law.cornell.edu/cfr/text/48/part-239/subpart-239.73

Code of Federal Regulations, Title 48, Federal Acquisition Regulations System, Section 252.239-7017, Notice of Supply Chain Risk. As of August 6, 2020: https://www.law.cornell.edu/cfr/text/48/252.239-7017

Code of Federal Regulations, Title 48, Federal Acquisition Regulations System, Section 252.239-7018, Supply Chain Risk. As of August 6, 2020: https://www.law.cornell.edu/cfr/text/48/252.204-7018

Code of Federal Regulations, Title 48, Federal Acquisition Regulations System, Section 252.246-7007, Contractor Counterfeit Electronic Part Detection and Avoidance

System. As of August 13, 2020:
https://www.law.cornell.edu/cfr/text/48/252.246-7007

Cohen, William W., Pradeep Ravikumar, and Stephen E. Fienberg, "A Comparison of String Distance Metrics for Name-Matching Tasks," paper, American Association for Artificial Intelligence, 2003. As of August 12, 2020:
http://dc-pubs.dbs.uni-leipzig.de/files/Cohen2003Acomparisonofstringdistance.pdf

"Data Matching Part 3: Match Scoring," *DS lore* blog, July 29, 2016. As of August 5, 2020:
http://nadbordrozd.github.io/blog/2016/07/29/datamatching-part-3-match-scoring/

Defense Contract Management Agency, "Defense Priorities & Allocations System (DPAS)," webpage, May 7, 2019. As of August 3, 2020:
https://www.dcma.mil/DPAS/

Defeo, Joseph A., ed., *Juran's Quality Handbook*, 7th ed., New York: McGraw-Hill, 2017.

Defense Intelligence Agency Instruction 5000.002, *Intelligence Threat Support for Major Defense Acquisition Programs*, June 19, 2019.

Department of Defense Directive 5148.13, *Intelligence Oversight*, Washington, D.C.: Office of the Deputy Chief Management Officer of the Department of Defense, April 26, 2017. As of July 31, 2020:
https://www.esd.whs.mil/Portals/54/Documents/DD/issuances/dodd/514813_dodd_2017.pdf

Department of Defense Instruction 4140.01, *DoD Supply Chain Materiel Management Policy*, Washington, D.C.: Under Secretary of Defense for Acquisition and Sustainment, March 6, 2019. As of July 31, 2020:
https://www.esd.whs.mil/Portals/54/Documents/DD/issuances/dodi/414001p.pdf

Department of Defense Instruction 4140.67, *DoD Counterfeit Prevention Policy*, USD(A&S) April 26, 2013, change 3, March 6, 2020. As of August 12, 2020:
https://www.esd.whs.mil/Portals/54/Documents/DD/issuances/dodi/414067p.pdf?ver=2020-03-06-140334-540

Department of Defense Instruction 5200.39, *Critical Program Information (CPI) Identification and Protection Within Research, Development, Test, and Evaluation (RDT&E)*, Washington, D.C.: Under Secretary of Defense for Intelligence and Under Secretary of Defense for Research and Engineering, May 28, 2015, change 2, October 15, 2018. As of July 31, 2020:
https://www.esd.whs.mil/Portals/54/Documents/DD/issuances/dodi/520039p.pdf

Department of Defense Instruction 5200.44, *Protection of Mission Critical Functions to Achieve Trusted Systems and Networks (TSN)*, Washington, D.C.: DoD CIO/USD(R&E), July 27, 2017, change 3, October 15, 2018. As of July 31, 2020:
https://www.esd.whs.mil/Portals/54/Documents/DD/issuances/dodi/520044p.pdf

Department of Defense Instruction O-5240.24, *Counterintelligence Activities Supporting Research, Development, and Acquisition*, Washington, D.C.: Under Secretary of Defense for Intelligence, June 8, 2011, change 1, October 15, 2013, Not available to the general public.

Department of Defense Instruction 8500.01, *Cybersecurity*, Washington, D.C.: Department of Defense Chief Information Officer, March 14, 2014, change 1, October 7, 2019. As of August 12, 2020:
https://www.esd.whs.mil/portals/54/documents/dd/issuances/dodi/850001_2014.pdf

Department of Defense Manual 4140.01, *DoD Supply Chain Materiel Management Procedures* (in 12 volumes), various dates. As of August 26, 2020:
https://www.esd.whs.mil/Directives/issuances/dodm/

Department of Defense Manual 5240.01, *Procedures Governing the Conduct of DoD Intelligence Activities*, Washington, D.C.: Office of the Deputy Chief Management Officer of the Department of Defense, August 8, 2016. As of July 31, 2020:
https://www.esd.whs.mil/Portals/54/Documents/DD/issuances/dodm/
524001_dodm_2016.pdf?ver=2017-07-31-143413-363

Department of the Navy, Product Data Reporting and Evaluation Program (PDREP), "Who We Are," webpage, undated. As of August 3, 2020:
https://www.pdrep.csd.disa.mil/

DIA—*See* Defense Intelligence Agency.

Dillen, Paul, "And the Winner of Best FPGA of 2016 Is . . . ," *EE Times*, June 3, 2017. As of August 6, 2019:
https://www.eetimes.com/author.asp?doc_id=1331443

DoDI—*See* Department of Defense Instruction.

DoDM—*See* Department of Defense Manual.

DOJ—*See* U.S. Department of Justice.

ERAI, Inc., homepage, undated. As of August 3, 2020:
https://www.erai.com/

Executive Order 11858, *Foreign Investment in the United States*, Washington, D.C.: The White House, May 7, 1975. As of August 13, 2020:
https://www.archives.gov/federal-register/codification/executive-order/11858.html

Executive Order 12333, *United States Intelligence Activities*, Washington, D.C.: The White House, December 4, 1981. As of July 31, 2020:
https://www.archives.gov/federal-register/codification/executive-order/12333.html

Executive Order 12334, *President's Intelligence Oversight Board*, Washington, D.C.: The White House, December 4, 1981. As of January 25, 2021:
https://www.cia.gov/readingroom/docs/CIA-RDP85-00988R000200260002-1.pdf

Export Control Act of 2018—*See* Subtitle B, Part 1, Pub. L. 115-232.

FactSet website, undated. As of July 31, 2020:
https://www.factset.com/

Federal Acquisition Regulation, Part 15, Contracting by Negotiation. As of August 6, 2020:
https://www.acquisition.gov/content/part-15-contracting-negotiation

Fergusson, Ian F., and Paul K. Kerr, *The U.S. Export Control System and the Export Control Reform Initiative*, Washington, D.C.: Congressional Research Service, R41916, April 5, 2019. As of July 31, 2020:
https://www.everycrsreport.com/files/
20190405_R41916_6ee7c001ca2f98acd9de987e576d8fafec4b4971.pdf

FIRRMA—*See* Foreign Investment Risk Review Modernization Act of 2018 in Pub. L. 115-232.

Galvan, David A., Brett Hemenway, William Welser IV, and Dave Baiocchi, *Satellite Anomalies: Benefits of a Centralized Anomaly Database and Methods for Securely Sharing Information Among Satellite Operators*, Santa Monica, Calif.: RAND Corporation, RR-560-DARPA, 2014. As of September 17, 2019:
https://www.rand.org/pubs/research_reports/RR560.html

GAO—*See* Government Accountability Office.

Gao, Renata, "Honeywell and AVIC FACRI Hold Opening Ceremony for Honfei's New Plant," press release, Honeywell website, March 25, 2017. As of July 31, 2019:
https://aerospace.honeywell.com/en/press-release-listing/2018/
december/honeywell-and-avic-facri-hold-opening-ceremony-for-honfei-s-new-plant

Government Accountability Office, *Counterfeit Parts: DOD Needs to Improve Reporting and Oversight to Reduce Supply Chain Risk*, Washington, D.C., GAO-16-236, February 2016. As of January 25, 2021:
https://www.gao.gov/assets/680/675227.pdf

———, *Integrating Existing Supplier Data and Addressing Workforce Challenges Could Improve Risk Analysis,* Washington, D.C., GAO-18-435, 2018.

Greenberg, Andy, "Software Has a Serious Supply-Chain Security Problem," *Wired*, September 18, 2017. As of August 6, 2019:
https://www.wired.com/story/ccleaner-malware-supply-chain-software-security/

ICD—*See* Intelligence Community Directive.

ICS—*See* Intelligence Community Standard.

Intelligence Community Directive 731, *Supply Chain Risk Management*, Office of the Director of National Intelligence, December 7, 2013. As of July 31, 2020:
https://www.dni.gov/files/NCSC/documents/supplychain/
20190327-ICD731-Supply-Chain-Risk-Manage20131207.pdf

Intelligence Community Standard 731-02, *Supply Chain Threat Assessments*, Office of the Director of National Intelligence, May 17, 2016.

Interagency Task Force, "Assessing and Strengthening the Manufacturing and Defense Industrial Base and Supply Chain Resiliency of the United States," Washington, D.C.: Office of the Undersecretary of Defense for Acquisition and Sustainment and Office of the Deputy Assistant Secretary of Defense for Industrial Policy, September 2018, As of July 11, 2019:
https://media.defense.gov/2018/Oct/05/2002048904/-1/-1/1/
assessing-and-strengthening-the-manufacturing-and
%20defense-industrial-base-and-supply-chain-resiliency.pdf

Jackson, James K., *The Committee on Foreign Investment in the United States (CFIUS)*, Washington, D.C.: Congressional Research Service, RL33388, August 6, 2019. As of July 31, 2020:
https://crsreports.congress.gov/product/pdf/RL/RL33388/81

Juran, J. M., and Frank M. Gryna, eds., *Juran's Quality Control Handbook*, New York: McGraw-Hill, 1988.

Lohr, Steve, "State Department Yields on PC's from China," *New York Times*, May 23, 2006.

Macaluso, Mark, "Honeywell to Invest in Leading Chinese Supply Chain Software Provider and Form a New Joint Venture to Pursue Opportunities Outside China," press release, December 7, 2017. As of July 31, 2019:
https://www.honeywell.com/content/honeywell/us/en/newsroom/pressreleases/2017/12/
honeywell-to-invest-in-leading-chinese-supply-chain-software-provider-and-form-a-new-
joint-venture-to-pursue-opportunities-outside-china

Mandatory Procedure 5315.3, "Source Selection," *Air Force Federal Acquisition Regulation Supplement*, 2019. As of August 13, 2020:
https://www.acquisition.gov/affars/source-selection

Marchese, Kelly, and Siva Paramasivam, "The Ripple Effect: How Manufacturing and Retail Executives View the Growing Challenge of Supply Chain Risk," New York: Deloitte Development LLC, 2013. As of July 19, 2019:
https://www2.deloitte.com/content/dam/Deloitte/global/Documents/
Process-and-Operations/gx-operations-consulting-the-ripple-effect-041213.pdf

MarketWatch, "Global Embedded Field-Programmable Gate Array (FPGA) Market by 2023: Global Industry Report with Manufacturers, Regions, Trends, Challenges, Market Size, Product Types and Applications," press release, June 30, 2020. As of August 12, 2020: https://www.marketwatch.com/press-release/global-embedded-field-programmable-gate-array-fpga-market-by-2023-global-industry-report-with-manufacturers-regions-trends-challenges-market-size-product-types-and-applications-2020-06-30

Mayer, Lauren A., Mark V. Arena, Frank Camm, Jonathan P. Wong, Gabriel Lesnick, Sarah Soliman, Edward Fernandez, Phillip Carter, and Gordon T. Lee, *Prototyping Using Other Transactions: Case Studies for the Acquisition Community,* Santa Monica, Calif.: RAND Corporation, RR-4417-AF, 2020. As of January 25, 2021: https://www.rand.org/pubs/research_reports/RR4417.html

McCarthy, Kerry R., Matthew R. Peterson, Jennifer J. Shafer, Jennifer Bisceglie, Dan Colman, and Brent Wildasin, *DoD Supply Chain Risk Management: Assessment and Recommendations: Assessment and Recommendations*, Tysons, Va.: LMI, March 2018.

Miller, Jason, "Why the Navy Is Giving Agencies, Industry a Much-Needed Wake-Up Call on Supply Chain Risks," Federal News Network website, April 4, 2019. As of August 6, 2019: https://federalnewsnetwork.com/acquisition/2019/04/navy-giving-agencies-industry-much-needed-wake-up-call-on-supply-chain-risks/

Moore, Nancy Y., and Elvira N. Loredo, *Identifying and Managing Air Force Sustainment Supply Chain Risks*, Santa Monica, Calif.: RAND Corporation, DB-649-AF, 2013. As of May 16, 2019: https://www.rand.org/pubs/documented_briefings/DB649.html

Moore, Nancy Y., Clifford A. Grammich, and Judith D. Mele, *Findings from Existing Data on the Department of Defense Industrial Base*, Santa Monica, Calif.: RAND Corporation, RR-614-OSD, 2014. As of July 18, 2019: https://www.rand.org/pubs/research_reports/RR614.html

National Institute of Standards and Technology, "NIST Roadmap for Improving Critical Infrastructure Cybersecurity," February 12, 2014. As of August 3, 2020: https://www.nist.gov/system/files/roadmap-021214.pdf

———, "Cyber Supply Chain Risk Management," June 22, 2020. As of August 12, 2020. https://csrc.nist.gov/projects/cyber-supply-chain-risk-management/key-practices

"Natural Language Toolkit," NLTK 3.5 Documentation website, April 13, 2020. As of August 12, 2020: https://www.nltk.org

Nissen, Chris, John Gronager, Robert Metzger, and Harvey Rishikof, *Deliver Uncompromised: A Strategy for Supply Chain Security and Resilience in Response to the Changing Character of War*, McLean, Va.: MITRE Center for Technology & National Security, August 2018.

NIST—*See* National Institute of Standards and Technology.

Nimbis Services Incorporated, "Trusted Silicon Stratus (TSS) Workshop," McLean, Va., February 2011. As of August 3, 2020:
https://apps.dtic.mil/dtic/tr/fulltext/u2/a540791.pdf

Office of Strategic Services, *Simple Sabotage Field Manual*, Washington, D.C., January 17, 1944. As of August 3, 2020:
https://www.cia.gov/news-information/featured-story-archive/
2012-featured-story-archive/CleanedUOSSSimpleSabotage_sm.pdf

Office of the United States Trade Representative, *2018 Special 301 Report*, Washington, D.C., 2018. As of August 3, 2020.
https://ustr.gov/sites/default/files/files/Press/Reports/2018%20Special%20301.pdf

Public Law 73-291, Securities Exchange Act of 1934, June 6, 1934. As of August 11, 2020:
https://fraser.stlouisfed.org/files/docs/historical/congressional/
securities-exchange-act.pdf

Public Law 83-703, Atomic Energy Act of 1954, August 30, 1954. As of August 11, 2020:
https://www.nrc.gov/docs/ML1327/ML13274A489.pdf

Public Law 94–329, International Security Assistance and Arms Export Control Act of 1976, June 30, 1976. As of August 11, 2020:
https://www.govinfo.gov/content/pkg/STATUTE-90/pdf/STATUTE-90-Pg729.pdf

Public Law 95-213, Foreign Corrupt Practices Act of 1977, December 19, 1977. As of August 6, 2020:
https://www.govinfo.gov/content/pkg/STATUTE-91/pdf/STATUTE-91-Pg1494.pdf

Public Law 95–223, With Respect to the Powers of the President in Time of War or National Emergency, October 28, 1977. As of August 11, 2020:
https://www.govinfo.gov/content/pkg/STATUTE-91/pdf/STATUTE-91-Pg1625.pdf#page=2

Public Law 98-369, Deficit Reduction Act of 1984, July 18, 1984. As of August 11, 2020:
https://www.govinfo.gov/content/pkg/STATUTE-98/pdf/STATUTE-98-Pg494.pdf

Public Law 112–81, National Defense Authorization Act for Fiscal Year 2012, December 31, 2011. As of August 3, 2020:
https://www.govinfo.gov/content/pkg/PLAW-112publ81/pdf/PLAW-112publ81.pdf

Public Law 115-91, National Defense Authorization Act for Fiscal Year 2018, December 12, 2017. As of August 4, 2020:
https://www.congress.gov/115/plaws/publ91/PLAW-115publ91.pdf

Public Law 115-232, John S. McCain National Defense Authorization Act for Fiscal Year 2019, August 13, 2018. As of August 4, 2020:
https://www.congress.gov/115/plaws/publ232/PLAW-115publ232.pdf

Public Law 115-390, Strengthening and Enhancing Cyber-Capabilities by Utilizing Risk Exposure Technology Act (SECURE Technology Act), December 21, 2018. As of August 3, 2020:
https://www.congress.gov/115/plaws/publ390/PLAW-115publ390.pdf

Robertson, Stephen, "Understanding Inverse Document Frequency: On Theoretical Arguments for IDF," *Journal of Documentation*, Vol. 60, No. 5, October 2004, pp. 503–520. As of August 4, 2020:
http://citeseerx.ist.psu.edu/viewdoc/download?doi=10.1.1.438.2284&rep=rep1&type=pdf

Rockwell Collins, Inc., Securities and Exchange Commission Form 10-K for the Fiscal Year Ended September 30, 2018, Cedar Rapids, Ia., November 26, 2018. As of July 31, 2019:
https://www.sec.gov/Archives/edgar/data/1137411/000113741118000111/col_9302018x10k.htm

S&P Global Market Intelligence website, undated. As of July 31, 2020:
https://www.spglobal.com/marketintelligence/en/solutions/sp-capital-iq-platform

Schmidt, William, and David Simchi-Levi, "Nissan Motor Company LTC.: Building Operational Resiliency," Cambridge, Mass.: MIT Sloan Management, August 27, 2013. As of July 19, 2019:
https://mitsloan.mit.edu/LearningEdge/CaseDocs/13-149%20Nissan.Simchi-Levi.pdf

SCRM WG—*See* Air Force Supply Chain Risk Management Working Group.

SEC—*See* U.S. Securities and Exchange Commission.

Shanahan, Raymond C., "Field Programmable Gate Array (FPGA) Assurance," presentation at the 20th Annual NDIA Systems Engineering Conference Springfield, Va., October 26, 2017.

Sheffi, Yossi, *The Power of Resilience: How the Best Companies Manage the Unexpected*, Cambridge, Mass.: MIT Press, 2015.

Trimberger, Stephen M., and Jason J. Moore. "FPGA Security: Motivations, Features, and Applications," *Proceedings of the IEEE*, Vol. 102, No. 8, August 2014, pp. 1248–1265. As of August 3, 2020:
https://ieeexplore.ieee.org/document/6849432

USA Spending, "Custom Award Data," webpage, undated a. As of August 3, 2020:
https://www.usaspending.gov/#/download_center/custom_award_data

USA Spending, "Data Dictionary," webpage, undated b. As of August 6, 2020:
https://www.usaspending.gov/#/download_center/data_dictionary

U.S.C.—*See* U.S. Code.

U.S. Code, Title 10, Armed Forces, Section 2304, Contracts: Competition Requirements.

U.S. Code, Title 10, Armed Forces, Section 2305, Contracts: Planning, Solicitation, Evaluation, and Award Procedure.

U.S. Code, Title 10, Armed Forces, Section 2339a, Requirements for Information Relating to Supply Chain Risk.

U.S. Code, Title 41, Public Contracts, Section 1322, Federal Acquisition Security Council Establishment and Membership.

U.S. Code, Title 50, War and National Defense, Appendix, Section 2170, Authority to Review Certain Mergers, Acquisitions, and Takeovers, Subsection (g) Additional Information to Congress; Confidentiality.

U.S. Code, Title 50, War and National Defense, Section 4565, Authority to Review Certain Mergers, Acquisitions, and Takeovers.

U.S. Department of Commerce, Bureau of Industry and Security, annual report archive, 2009. As of July 18, 2019:
https://www.bis.doc.gov/index.php/about-bis/newsroom/archives/annual-reports

U.S. Department of Defense, Office of Small Business Programs, "Small Business Program Goals and Performance," webpage, 2020. As of July 31, 2019:
https://business.defense.gov/About/Goals-and-Performance/

U.S. Department of Justice, "Enforcement Actions," webpage, July 2, 2020. As of July 18, 2019:
https://www.justice.gov/criminal-fraud/related-enforcement-actions

U.S. Department of State, Directorate of Defense Trade Controls, "Penalties & Oversight Agreements," webpage, 2020. As of July 18, 2019:
https://www.pmddtc.state.gov/
ddtc_public?id=ddtc_kb_article_page&sys_id=384b968adb3cd30044f9ff621f961941

U.S. Securities and Exchange Commission, "EDGAR: Company Filings." webpage, undated. As of August 3, 2020:
https://www.sec.gov/edgar/searchedgar/companysearch.html

———, "Form 10-K," webpage, June 26, 2009. As of August 3, 2020:
https://www.sec.gov/fast-answers/answers-form10khtm.html

———, "How to Read a 10-K," webpage, July 21, 2011. As of August 5, 2020:
https://www.sec.gov/fast-answers/answersreada10khtm.html

———, "Smaller Reporting Company Definition," Washington, D.C., June 28, 2018. As of December 17, 2019:
https://www.sec.gov/rules/final/2018/33-10513.pdf

———, "Amendments to the Smaller Reporting Company Definition: A Small Entity Compliance Guide for Issuers," webpage, August 10, 2018. As of August 12, 2020:
https://www.sec.gov/corpfin/amendments-smaller-reporting-company-definition

———, "Exchange Act Reporting and Registration," webpage, October 24, 2018. As of August 5, 2020:
https://www.sec.gov/smallbusiness/goingpublic/exchangeactreporting

———, "SEC Enforcement Actions: FCPA Cases," webpage, May 13, 2019. As of July 18, 2019:
https://www.sec.gov/spotlight/fcpa/fcpa-cases.shtml

———, "Smaller Reporting Companies," webpage, July 24, 2019. As of December 17, 2019:
https://www.sec.gov/smallbusiness/goingpublic/SRC

Wall Street Prep, Inc., "Bloomberg vs. Capital IQ vs. FactSet vs. Thomson Reuters Eikon," webpage, undated. As of December 19, 2019:
https://www.wallstreetprep.com/knowledge/bloomberg-vs-capital-iq-vs-factset-vs-thomson-reuters-eikon/

"What Is the Difference Between FPDS-NG and USA Spending.gov?" FPDS-NG FAQ, undated. As of August 12, 2020:
https://www.fpds.gov/wiki/index.php/FPDS-NG_FAQ#What.E2.80.99s_the_difference_between_FPDS-NG_and_USASpending.gov.3F

CPSIA information can be obtained
at www.ICGtesting.com
Printed in the USA
LVHW061956141021
700457LV00012B/396

9 781977 406583